Military Rigging Handbook

edited by
Brian Greul

This manual is a guide and basic reference for personnel whose duties require the use of rigging. It is intended for use in training and as a reference manual for field operations. It covers the types of rigging and the application of fiber rope, wire rope, and chains used in various combinations to raise or move heavy loads. It includes basic instructions on knots, hitches, splices, lashing, and tackle systems. Safety precautions and requirements for the various operations are listed, as well as rules of thumb for rapid safe-load calculations.

Should you have suggestions or feedback on ways to improve this book please send email to Books@OcotilloPress.com

Edited 2021 Ocotillo Press
ISBN 978-1-954285-34-7

Printed in the United States of America

Ocotillo Press
Houston, TX 77017
Books@OcotilloPress.com

FM 5-125

Rigging Techniques, Procedures, and Applications

HEADQUARTERS, DEPARTMENT OF THE ARMY

DISTRIBUTION RESTRICTION: *Approved for public release; distribution is unlimited.*

Change 1

Headquarters
Department of the Army
Washington, DC, 23 February 2001

Rigging Techniques, Procedures, and Applications

1. Change FM 5-125, 3 October 1995, as follows:

Remove Old Pages	Insert New Pages
i through iv	i through iv
vii and viii	vii and viii
1-15 through 1-20	1-15 through 1-21
2-37 through 2-40	2-37 through 2-40

2. A bar (|) marks new or changed material.

3. File this transmittal sheet in front of the publication.

DISTRIBUTION RESTRICTION: Approved for public release; distribution is unlimited.

ERIC K. SHINSEKI
General, United States Army
Chief of Staff

Official:

JOEL B. HUDSON
Adiministrative Assistant to the
Secretary of the Army
0100405

DISTRIBUTION:

Active Army, Army National Guard, and US Army Reserve: To be distributed in accordance with the initial distribution number 115426, requirements for FM 5-125.

*FM 5-125

Field Manual
No. 5-125

Headquarters
Department of the Army
Washington, DC, 3 October 1995

Rigging Techniques, Procedures, and Applications

Contents

*This manual supersedes TM 5-725, 3 October 1968.

ListofFigures

List of Tables

Preface

This manual is a guide and basic reference for personnel whose duties require the use of rigging. It is intended for use in training and as a reference manual for field operations. It covers the types of rigging and the application of fiber rope, wire rope, and chains used in various combinations to raise or move heavy loads. It includes basic instructions on knots, hitches, splices, lashing, and tackle systems. Safety precautions and requirements for the various operations are listed, as well as rules of thumb for rapid safe-load calculations.

The material contained herein is applicable to both nuclear and nonnuclear warfare.

The proponent for this publication is Headquarters (HQ), United States (US) Army Training and Doctrine Command (TRADOC). Users of this manual are encouraged to submit recommended changes or comments on Department of the Army (DA) Form 2028 and forward them to: Commandant, US Army Engineer School, ATTN: ATSE-T-PD-P, Fort Leonard Wood, Missouri 65473-6500.

Unless otherwise stated, masculine nouns and pronouns do not refer exclusively to men.

CHAPTER 1

Rope

Section 1. Fiber Rope

In the fabrication of fiber rope, a number of fibers of various plants are twisted together to form yarns. These yarns are then twisted together in the opposite direction of the fibers to form strands (see *Figure 1-1, page 1-2)* The strands are twisted in the opposite direction of the yarns to form the completed rope. The direction of twist of each element of the rope is known as the "lay" of that element. Twisting each element in the opposite direction puts the rope in balance and prevents its elements from unlaying when a load is suspended on it. The principal type of fiber rope is the three-strand, right lay, in which three strands are twisted in a right-hand direction. Four-strand ropes, which are also available, are slightly heavier but are weaker than three-strand ropes of the same diameter.

TYPES OF FIBERS

The term cordage is applied collectively to ropes and twines made by twisting together vegetable or synthetic fibers.

VEGETABLE FIBERS

The principal vegetable fibers are abaca (known as Manila), sisalana and henequen (both known as sisal), hemp, and sometimes coir, cotton, and jute. The last three are relatively unimportant in the heavy cordage field.

Abaca, sisalana, and henequen are classified as hard fibers. The comparative strengths of the vegetable fibers, considering abaca as 100, are as follows:

- Sisalana   80
- Henequen   65
- Hemp   100

Manila

This is a strong fiber that comes from the leaf stems of the stalk of the abaca plant, which belongs to the banana family. The fibers vary in length from 1.2 to 4.5 meters (4 to 15 feet) in the natural states. The quality of the fiber and its length give Manila rope relatively high elasticity, strength, and resistance to wear and deterioration. The manufacturer treats the rope with chemicals to make it more mildew resistant, which increases the rope's quality. Manila rope is generally the standard item of issue because of its quality and relative strength.

Sisal

Sisal rope is made from two tropical plants, sisalana and henequen, that produce fibers 0.6 to 1.2 meters (2 to 4 feet)

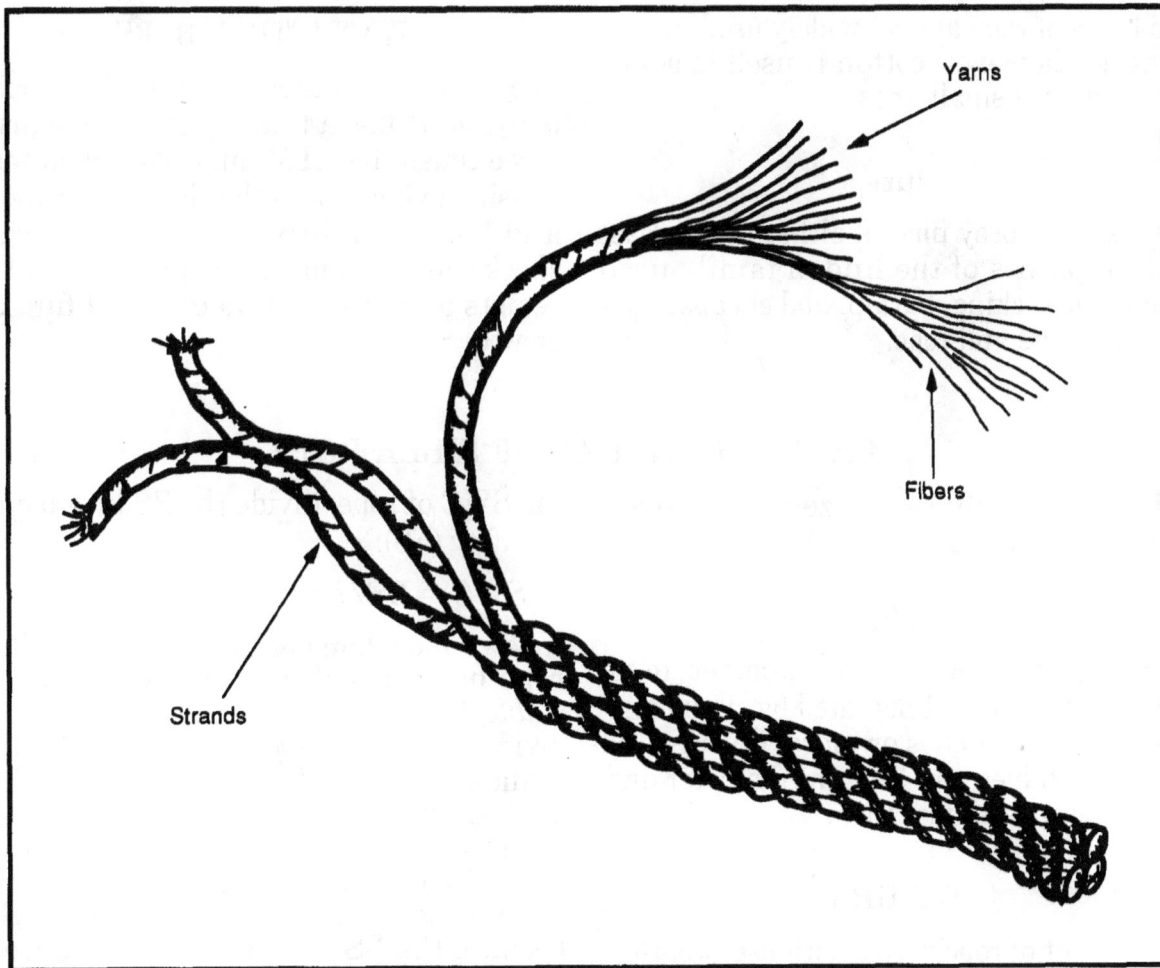

Figure 1-1. Cordage of rope construction

long. Sisalana produces the stronger fibers of the two plants, so the rope is known as sisal. Sisal rope is about 80 percent as strong as high quality Manila rope and can be easily obtained. It withstands exposure to sea water very well and is often used for this reason.

Hemp

This tall plant is cultivated in many parts of the world and provides useful fibers for making rope and cloth. Hemp was used extensively before the introduction of Manila, but its principal use today is in fittings, such as ratline, marline, and spun yarn. Since hemp absorbs much better than the hard fibers, these fittings are invariably tarred to make them more water-resistant. Tarred hemp has about 80 percent of the strength of untarred hemp. Of these tarred fittings, marline is the standard item of issue.

Coir and Cotton

Coir rope is made from the fiber of coconut husks. It is a very elastic, rough rope about one-fourth the strength of hemp but light enough to float on water. Cotton makes a very smooth white rope that withstands much bending and running. These two types of rope are not widely used in the military; however, cotton is used in some cases for very small lines.

Jute

Jute is the glossy fiber of either of two East Indian plants of the linden family used chiefly for sacking, burlap, and cheaper varieties of twine and rope.

SYNTHETIC FIBERS

The principal synthetic fiber used for rope is nylon. It has a tensile strength nearly three times that of Manila. The advantage of using nylon rope is that it is waterproof and has the ability to stretch, absorb shocks, and resume normal length. It also resists abrasion, rot, decay, and fungus growth.

CHARACTERISTICS OF FIBER ROPE

Fiber rope is characterized by its size, weight, and strength.

SIZE

Fiber rope is designated by diameter up to 5/8 inch, then it is designated by circumference up to 12 inches or more. For this reason, most tables give both the diameter and circumference of fiber rope.

WEIGHT

The weight of rope varies with use, weather conditions, added preservatives, and other factors. *Table 1-1, page 1-4*, lists the weight of new fiber rope.

STRENGTH

Table 1-1 lists some of the properties of Manila and sisal rope, including the breaking strength (B S), which is the greatest stress that a material is capable of withstanding without rupture. The table shows that the minimum BS is considerably greater than the safe load or the safe working capacity (SWC). This is the maximum load that can safely be applied to a particular type of rope. The difference is caused by the application of a safety factor. To obtain

the SWC of rope, divide the BS by a factor of safety (FS):

$$SWC = BS/FS$$

A new l-inch diameter, Number 1 Manila rope has a BS of 9,000 pounds (see *Table 1-1*).To determine the rope's SWC, divide its BS (9,000 pounds) by a minimum standard FS of 4. The result is a SWC of *2,250* pounds. This means that you can safely apply 2,250 pounds of tension to the new l-inch diameter, Number 1 Manila rope in normal use. Always use a FS because the BS of rope becomes reduced after use and exposure to weather conditions. In addition, a FS is required because of shock loading, knots, sharp bends, and other stresses that rope may have to withstand during its use. Some of these stresses reduce the strength of rope as much as 50 percent. If tables are not available, you can closely approximate the SWC by a rule of thumb. The rule of thumb for the SWC, in tons, for fiber rope is equal to the square of the rope diameter (D) in inches:

$$SWC = D^2$$

The SWC, in tons, of a l/2-inch diameter fiber rope would be 1/2 inch squared or 1/4 ton. The rule of thumb allows a FS of about 4.

Table 1-1. Properties of manila and sisal rope

Nominal Diameter (inches) Circumference (inches) Pounds per Foot			Number 1 Manila		Sisal	
			Breaking Strength (pounds)	Safe Load (pounds) FS = 4	Breaking Strength (pounds)	Safe Load (pounds) FS = 4
1/4	3/4	0.020	600	150	480	120
3/8	1 1/8	0.040	1,350	325	1,080	260
1/2	1 1/2	0.075	2,650	660	2,120	520
5/8	2	0.133	4,400	1,100	3,520	880
3/4	2 1/4	0.167	5,400	1,350	4,320	1,080
7/8	2 3/4	0.186	7,700	1,920	6,160	1,540
1	3	0.270	9,000	2,250	7,200	1,800
1 1/8	3 1/2	0.360	12,000	3,000	9,600	2,400
1 1/4	3 3/4	0.418	13,500	3,380	10,800	2,700
1 1/2	4 1/2	0.600	18,500	4,620	14,800	3,700
1 3/4	5 1/2	0.895	26,500	6,625	21,200	5,300
2	6	1.080	31,000	7,750	24,800	6,200
2 1/2	7 1/2	1.350	46,500	11,620	37,200	9,300
3	9	2.420	64,000	16,000	51,200	12,800

NOTES:
1. Breaking strengths and safe loads given are for new rope used under favorable conditions. As rope ages or deteriorates, reduce safe loads progressively to one-half of values given.
2. Safe working load may be computed using a safety factor of 4, but when the condition of the rope is doubtful, divide the computed further load by 2.

CARE OF FIBER ROPE

The strength and useful life of fiber rope is shortened considerably by improper care. To prolong its life and strength, observe the following guidelines:

- Ensure that it is dry and then stored in a cool, dry place. This reduces the possibility of mildew and rotting.

- Coil it on a spool or hang it from pegs in a way that allows air circulation.

- Avoid dragging it through sand or dirt or pulling it over sharp edges. Sand or grit between the fibers cuts them and reduces the rope's strength.

- Slacken taut lines before they are exposed to rain or dampness because a wet rope shrinks and may break.

- Thaw a frozen rope completely before using it; otherwise the frozen fibers will break as they resist bending.

- Avoid exposure to excessive heat and fumes of chemicals; heat or boiling water decreases rope strength about 20 percent.

HANDLING OF FIBER ROPE

New rope is coiled, bound, and wrapped in burlap. The protective covering should not be removed until the rope is to be used. This protects it during storage and prevents tangling. To open the new rope, strip off the burlap wrapping and look inside the coil for the end of the rope. This should be at the bottom of the coil (see *Figure 1-2*)If it is not, turn the coil over so the end is at the bottom. Pull the end up through the center of the coil. As the rope comes up, it unwinds in a counterclockwise direction.

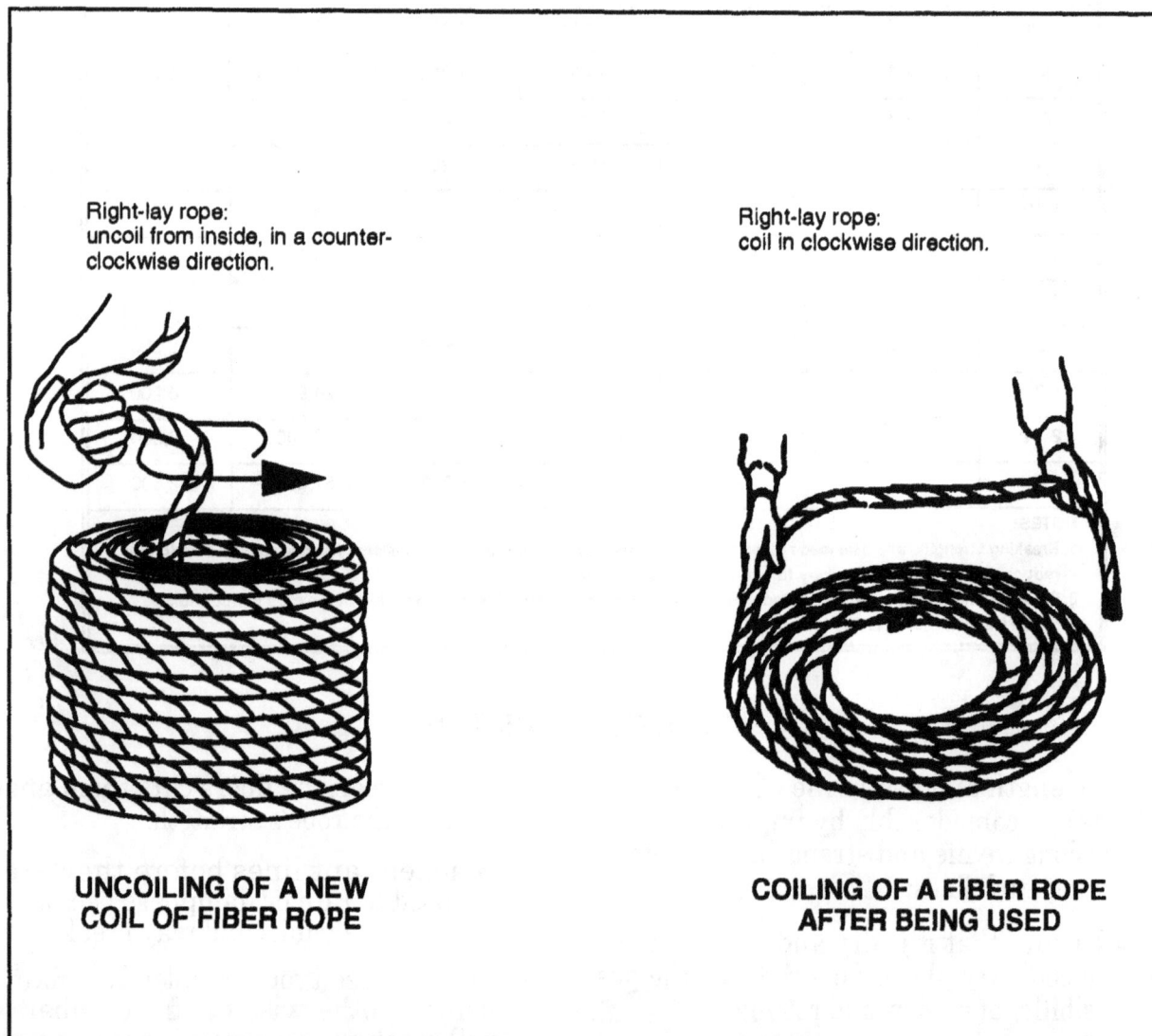

Right-lay rope: uncoil from inside, in a counter-clockwise direction.

Right-lay rope: coil in clockwise direction.

UNCOILING OF A NEW COIL OF FIBER ROPE

COILING OF A FIBER ROPE AFTER BEING USED

Figure 1-2. Uncoiling and coiling rope

INSPECTION OF FIBER ROPE

The outside appearance of fiber rope is not always a good indication of its internal condition. Rope softens with use. Dampness, heavy strain, fraying and breaking of strands, and chafing on rough edges all weaken it considerably. Overloading rope may cause it to break, with possible heavy damage to material and serious injury to personnel. For this reason, inspect it carefully at regular intervals to determine its condition. Untwist the strands slightly to open a rope so that you can examine the inside. Mildewed rope has a musty odor and the inner fibers of the strands have a dark, stained appearance. Broken strands or broken yarns ordinarily are easy to identify. Dirt and sawdust-like material inside a rope, caused by chafing, indicate damage. In rope having a central core, the core should not break away in small pieces when examined. If it does, this is an indication that a rope has been overstrained.

If a rope appears to be satisfactory in all other respects, pull out two fibers and try to break them. Sound fibers should offer considerable resistance to breakage. When you find unsatisfactory conditions, destroy a rope or cut it up in short pieces to prevent its being used in hoisting. You can use the short pieces for other purposes.

Section II. Wire Rope

The basic element of wire rope is the individual wire, which is made of steel or iron in various sizes. Wires are laid together to form strands, and strands are laid together to form rope (see *Figure 1-3*)Individual wires are usually wound or laid together in the opposite direction of the lay of the strands. Strands are then wound around a central core that supports and maintains the position of strands during bending and load stresses.

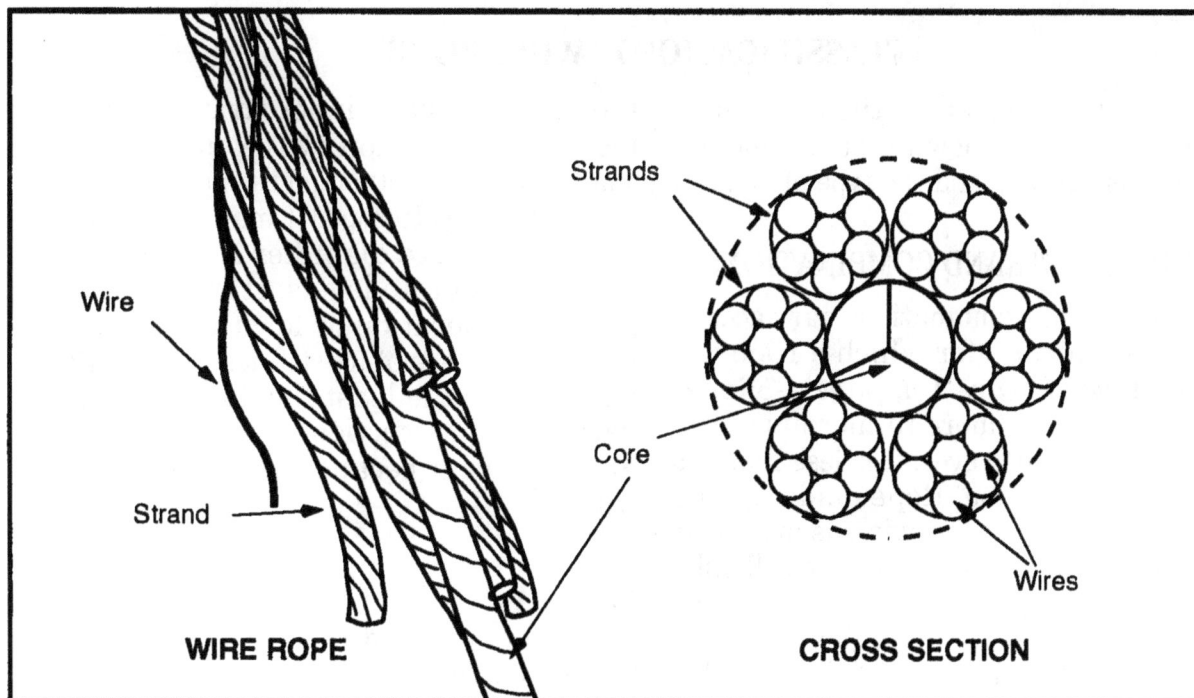

Figure 1-3. Elements of wire-rope construction

In some wire ropes, the wires and strands are preformed. Preforming is a method of presetting the wires in the strands (and the strands in the rope) into the permanent helical or corkscrew form they will have in the completed rope. As a result, preformed wire rope does not contain the internal stresses found in the nonpreformed wire rope; therefore, it does not untwist as easily and is more flexible than nonpreformed wire rope.

TYPES OF WIRE ROPE CORES

The core of wire rope may be constructed of fiber rope, independent wire rope, or a wire strand.

FIBER-ROPE CORES

The fiber-rope core can be of vegetable or synthetic fibers. It is treated with a special lubricant that helps keep wire rope lubricated internally. Under tension, wire rope contracts, forcing the lubricant from the core into the rope. This type of core also acts as a cushion for the strands when they are under stress, preventing internal crushing of individual wires. The limitations of fiber-rope cores are reached when pressure, such as crushing on the drum, results in the collapse of the core and distortion of the rope strand. Furthermore, if the rope is subjected to excessive heat, the vegetable or synthetic fibers may be damaged.

INDEPENDENT, WIRE-ROPE CORES

Under severe conditions, an independent, wire-rope core is normally used. This is actually a separate smaller wire rope that acts as a core and adds strength to the rope.

WIRE-STRAND CORES

A wire-strand core consists of a single strand that is of the same or a more flexible construction than the main rope strands.

CLASSIFICATION OF WIRE ROPE

Wire rope is classified by the number of strands, the number of wires per strand, the strand construction, and the type of lay.

WIRE AND STRAND COMBINATIONS

Wire and strand combinations vary according to the purpose for which a rope is intended (see *Figure 1-4, page 1-8*). Rope with smaller and more numerous wires is more flexible; however, it is less resistant to external abrasion. Rope made up of a smaller number of larger wires is more resistant to external abrasion but is less flexible. The 6-by-37 wire rope (6 strands, each made up of 37 wires) is the most flexible of the standard six-strand ropes. This flexibility allows it to be used with small drums and sheaves, such as on cranes. It is a very efficient rope because many inner strands are protected from abrasion by the outer strands. The stiffest and strongest type for general use is the 6-by-19 rope. It may be used over sheaves of large diameter if the speed is kept to moderate levels. It is not suitable for rapid operation or for use over small sheaves because of its stiffness. The 6-by-7 wire rope is the least flexible of the standard rope constructions. It can withstand abrasive wear because of the large outer wires.

LAY

Lay refers to the direction of winding of wires in strands and strands in rope (see *Figure 1-5, page 1-8*). Both may be wound in the same direction, or they may be wound in

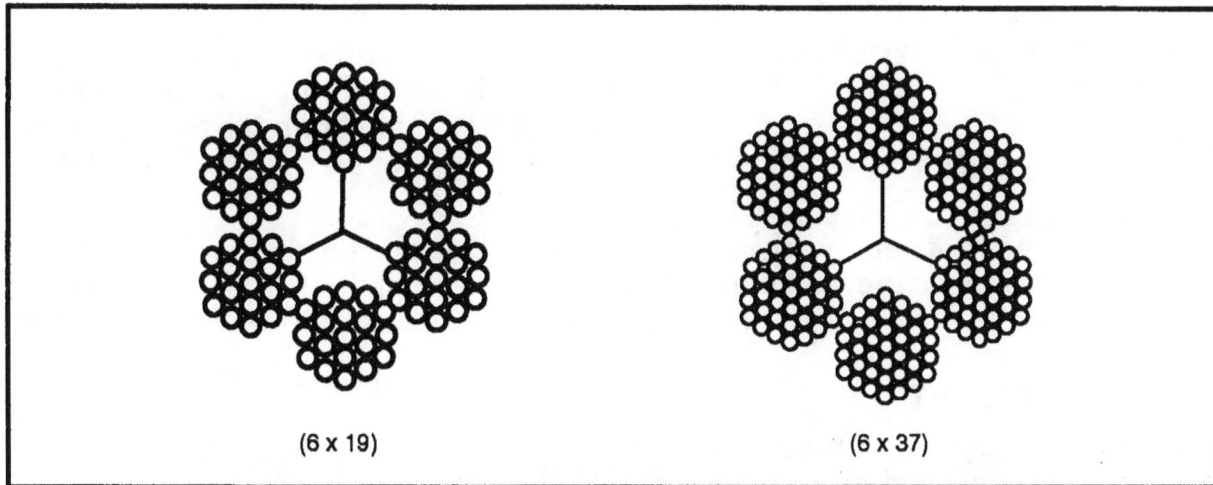

(6 x 19)   (6 x 37)

Figure 1-4. Arrangement of strands in wire rope

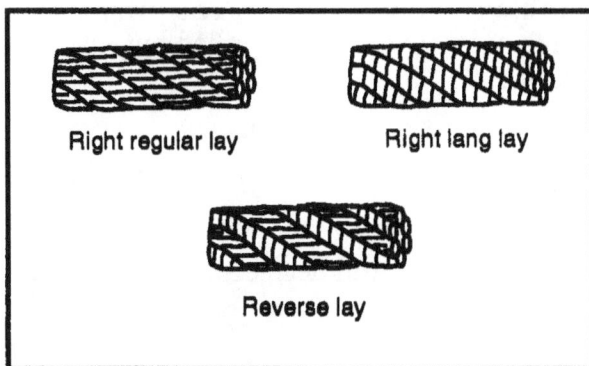

Right regular lay   Right lang lay

Reverse lay

Figure 1-5. Wire-rope lays

opposite directions. The three types of rope lays are—

- Regular.
- Lang.
- Reverse.

Regular Lay

In regular lay, strands and wires are wound in opposite directions. The most common lay in wire rope is right regular lay (strands wound right, wires wound left). Left regular lay (strands wound left, wires wound right) is used where the untwisting rotation of the rope counteracts the unscrewing forces in

the supported load, such as in drill rods and tubes for deep-well drilling.

Lang Lay

In lang lay, strands and wires are wound in the same direction. Because of the greater length of exposed wires, lang lay assures longer abrasion resistance of wires, less radial pressure on small diameter sheaves or drums by rope, and less binding stresses in wire than in regular lay wire rope. Disadvantages of lang lay are its tendencies to kink and unlay or open up the strands, which makes it undesirable for use where grit, dust, and moisture are present. The standard direction of lang lay is right (strands and wires wound right), although it also comes in left lay (strands and wires wound left).

Reverse Lay

In reverse lay, the wires of any strand are wound in the opposite direction of the wires in the adjacent strands. Reverse lay applies to ropes in which the strands are alternately regular lay and lang lay. The use of reverse lay rope is usually limited to certain types of conveyors. The standard direction of lay is right (strands wound right), as it is for both regular-lay and lang-lay ropes.

CHARACTERISTICS OF WIRE ROPE

Wire rope is characterized by its size, weight, and strength.

SIZE

The size of wire rope is designated by its diameter in inches. To determine the size of a wire rope, measure its greatest diameter (see *Figure 1-6*).

WEIGHT

The weight of wire rope varies with the size and the type of construction. No rule of thumb can be given for determining the weight. Approximate weights for certain sizes are given in *Table 1-2, page 1-10*.

STRENGTH

The strength of wire rope is determined by its size and grade and the method of fabrication. The individual wires may be made of various materials, including traction steel, mild plow steel (MPS), improved plow steel (IPS), and extra IPS. Since a suitable margin of safety must be provided when applying a load to a wire rope, the BS is divided by an appropriate FS to obtain the SWC for that particular type of service (see *Table 1-3, page 1-11*).

You should use the FS given in *Table 1-3* in all cases where rope will be in service for a considerable time. As a rule of thumb, you can square the diameter of wire rope in, inches, and multiply by 8 to obtain the SWC in tons:

$$SWC = 8D^2$$

A value obtained in this manner will not always agree with the FS given in *Table 1-3*. The table is more accurate. The proper FS depends not only on loads applied but also on the—

- Speed of the operation.
- Type of fittings used for securing the rope ends.
- Acceleration and deceleration.
- Length of the rope.

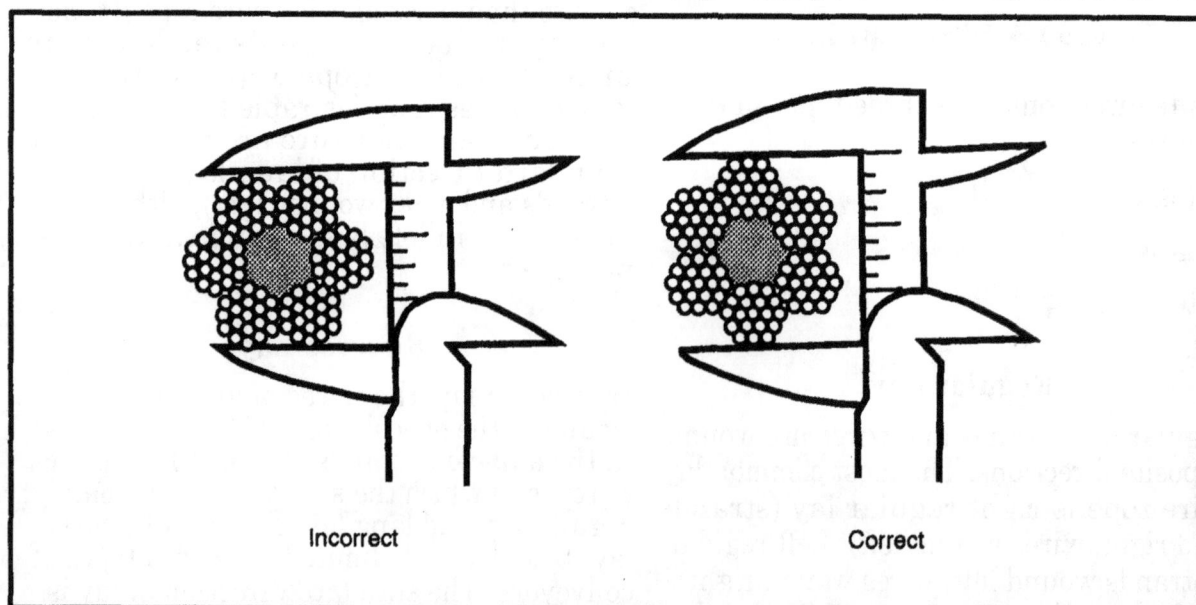

Incorrect   Correct

Figure 1-6. Measuring wire rope

Table 1-2. Breaking strength of 6 by 19 standard wire rope

Nominal Diameter (Inches) Approximate Weight (pounds per foot) Iron			Breaking Strength, Tons of 2,000 Pounds*			
			Traction Steel	Plow Steel	Improved Plow Steel	Extra Improved Plow Steel
1 1/2	3.60	29.7	56.6	108.92	123.43	123.43
1/4	0.10	1.4	2.6	2.39	2.74	
3/8	0.23	2.1	4.0	5.31	6.10	7.55
1/2	0.40	3.6	6.8	9.35	10.70	13.30
5/8	0.63	5.5	10.4	14.50	16.70	20.60
3/4	0.90	7.9	14.8	20.70	23.80	29.40
7/8	1.23	10.6	20.2	28.00	32.20	39.80
1	1.60	13.7	26.0	36.40	4.18	51.70
1 1/8	2.03	17.2	32.7	45.70	52.60	65.00
1 1/4	2.50	21.0	40.6	56.20	64.60	79.90
1 1/2	3.60	29.7	56.6	80.00	92.00	114.00
1 3/4				108.00	124.00	153.00
2				139.00	160.00	198.00

*The maximum allowable working load is the breaking strength divided by the appropriate factor of safety (*see Table 1-3*).

- Number, size, and location of sheaves and drums.
- Factors causing abrasion and corrosion.
- Facilities for inspection.
- Possible loss of life and property if the rope fails.

Table 1-2 shows comparative BS of typical wire ropes.

CARE OF WIRE ROPE

Caring for wire rope properly includes reversing the ends and cleaning, lubricating, and storing it. When working with wire rope, you should wear work gloves.

Table 1-3. Wire-rope FS

Type of Service	Minimum FS
Track cables	3.2
Guys	3.5
Miscellaneous hoisting equipment	5.0
Haulage ropes	6.0
Derricks	6.0
Small electric and air hoists	7.0
Slings	8.0

REVERSING OR CUTTING BACK ENDS

To obtain increased service from wire rope, it is sometimes advisable to either reverse or cut back the ends. Reversing the ends is more satisfactory because frequently the wear and fatigue on rope are more severe at certain points than at others. To reverse the ends, detach the drum end of the rope from the drum, remove the rope from the end attachment, and place the drum end of the rope in the end attachment. Then fasten the end that you removed from the end attachment to the drum. Cutting back the end has a similar effect, but there is not as much change involved. Cut a short length off the end of the rope and place the new end in the fitting, thus removing the section that has sustained the greatest local fatigue.

CLEANING

Scraping or steaming will remove most of the dirt or grit that may have accumulated on a used wire rope. Remove rust at regular intervals by using a wire brush. Always clean the rope carefully just before lubricating it. The object of cleaning at that time is to remove all foreign material and old lubricant from the valleys between the strands and from the spaces between the outer wires to permit the newly applied lubricant free entrance into the rope.

LUBRICATING

At the time of fabrication, a lubricant is applied to wire rope. However, this lubricant generally does not last throughout the life of the rope, which makes relubrication necessary. To lubricate, use a good grade of oil or grease. It should be free of acids and alkalis and should be light enough to penetrate between the wires and strands. Brush the lubricant on, or apply it by passing the rope through a trough or box containing the lubricant. Apply it as uniformly as possible throughout the length of the rope.

STORING

If wire rope is to be stored for any length of time, you should always clean and lubricate it first. If you apply the lubricant properly and store the wire in a place that is protected from the weather and from chemicals and fumes, corrosion will be virtually eliminated. Although the effects of rusting and corrosion of the wires and deterioration of the fiber core are difficult to estimate, it is certain that they will sharply decrease the strength of the rope. Before storing, coil the rope on a spool and tag it properly as to size and length.

HANDLING OF WIRE ROPE

Handling wire rope may involve kinking, coiling, unreeling, seizing, welding, cutting, *or* the use of drums and sheaves. When handling wire rope, you should wear work gloves.

KINKING

When handling loose wire rope, small loops frequently form in the slack portion (see *Figure 1-7*).If you apply tension while these loops are in position, they will not straighten out but will form sharp kinks, resulting in unlaying of the rope. You should straighten out all of these loops before applying a load. After a kink has formed in wire rope, it is impossible to remove it. Since the strength of the rope is seriously damaged at the point where a kink occurs, cut out that portion before using the rope again.

COILING

Small loops or twists will form if rope is being wound into the coil in a direction that is opposite to the lay. Coil left-lay wire rope in a counterclockwise direction and right-lay wire rope in a clockwise direction.

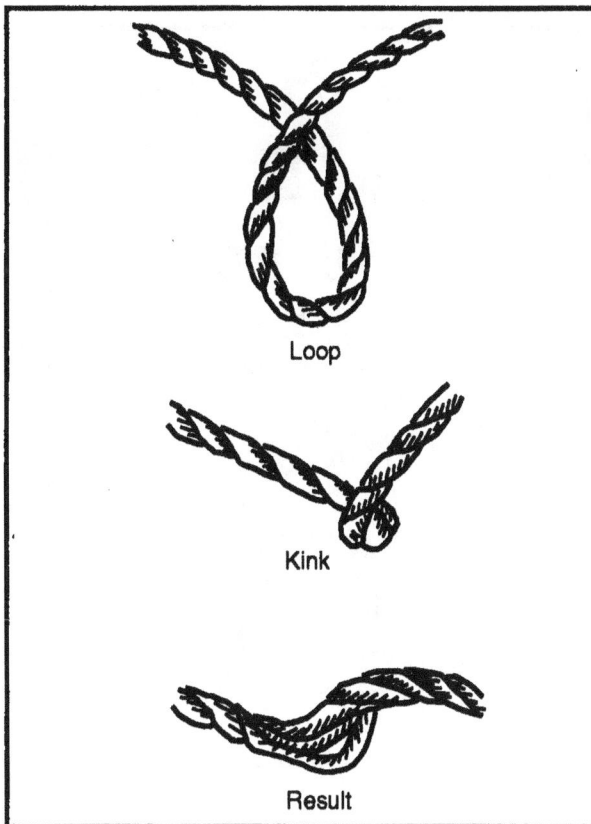

Figure 1-7. Kinking in wire rope

UNREELING

When removing wire rope from a reel or coil, it is imperative that the reel or coil rotate as the rope unwinds. Mount the reel as shown in *Figure 1-8*.Then pull the rope from the reel by holding the end of the rope and walking away from the reel, which rotates as the rope unwinds. If wire rope is in a small coil, stand the coil on end and roll it along the ground (see *Figure 1-9*)If loops form in the wire rope, carefully remove them before they form kinks.

SEIZING

Seizing is the most satisfactory method of binding the end of a wire rope, although welding will also hold the ends together satisfactorily. The seizing will last as long as desired, and there is no danger of weakening the wire through the application of heat. Wire rope is seized as shown in*Figure 1-10, page 1-14*.There are three convenient rules for determining the number of seizings, lengths, and space between seizings. In each case when the calculation results in a fraction, use the next larger whole number. The following calculations are based on a 4-inch diameter wire rope:

- *The number of seizings to be applied equals approximately three times the diameter of the rope (number of seizings = SD).*

- *Example: 3 x 3/4 (D) = 2 1/4. Use 3 seizings.*

- *Each seizing should be 1 to 1 1/2 times as long as the diameter of the rope. (length of seizing= 1 1/2D).*

- *The seizings should be spaced a distance apart equal to twice the diameter (spacing = 2D).*

Example: 2 x 3/4 (D) = 1 1/2. Use 2-inch spaces.

Figure 1-8. Unreeling wire rope

Figure 1-9. Uncoiling wire rope

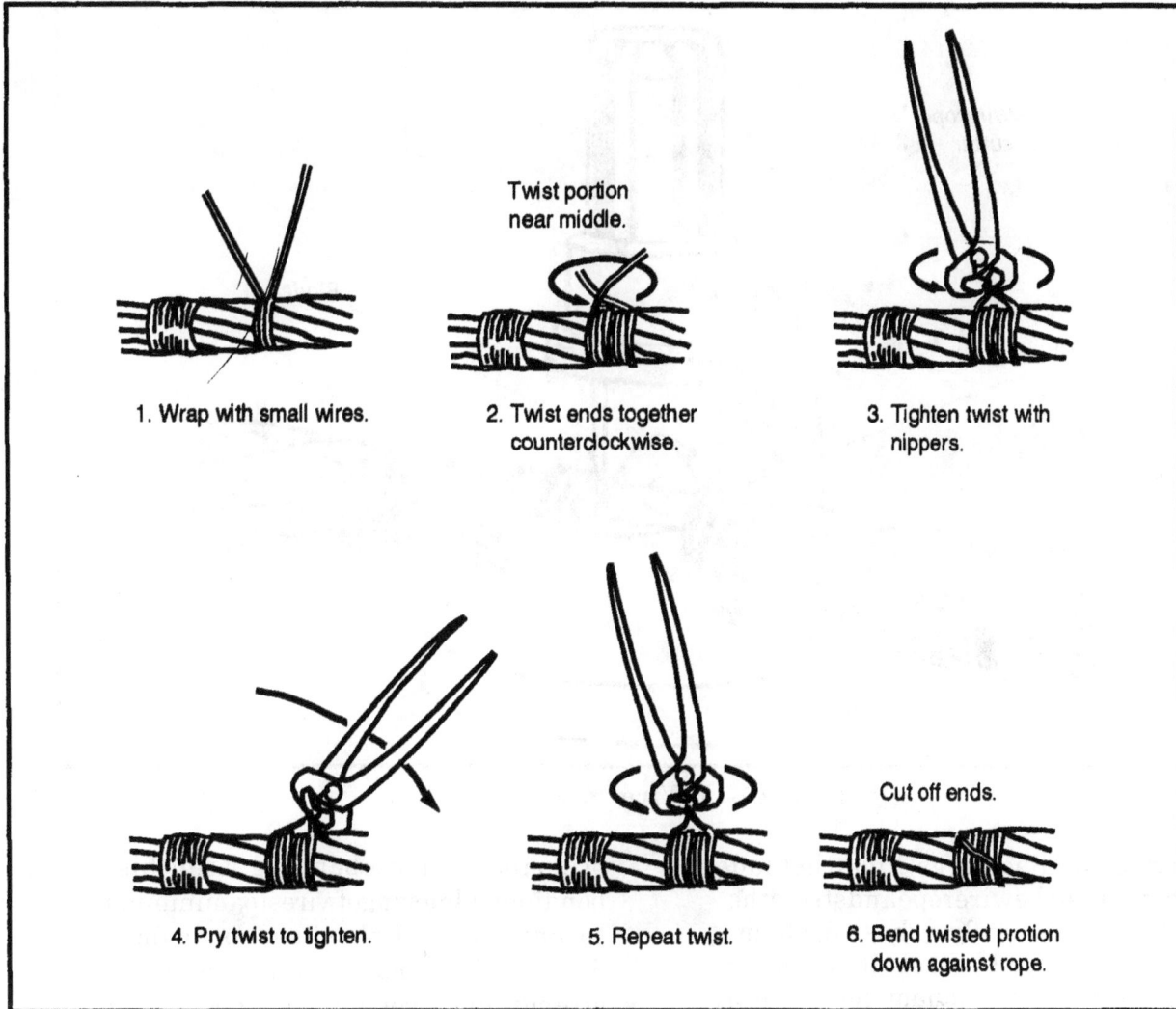

1. Wrap with small wires.

Twist portion near middle.

2. Twist ends together counterclockwise.

3. Tighten twist with nippers.

4. Pry twist to tighten.

5. Repeat twist.

Cut off ends.

6. Bend twisted protion down against rope.

Figure 1-10. Seizing wire rope

Note: Always change the fraction to the next larger whole number.

apply more heat than is essential to fuse the metal.

WELDING

You can bind wire-rope ends together by fusing or welding the wires. This is a satisfactory method if you do it carefully, as it does not increase the size of the rope and requires little time to complete. Before welding rope, cut a short piece of the core out of the end so that a clean weld will result and the core will not be burned deep into the rope. Keep the area heated to a minimum and do not

CUTTING

You can cut wire rope with a wire-rope cutter, a cold chisel, a hacksaw, bolt clippers, or an oxyacetylene cutting torch (see *Figure 1-11)*. Before cutting wire rope, tightly bind the strands to prevent unlaying. Secure the ends that are to be cut by seizing or welding them. To use the wire-rope cutter, insert the wire rope in the bottom of the cutter with the blade of the cutter coming between

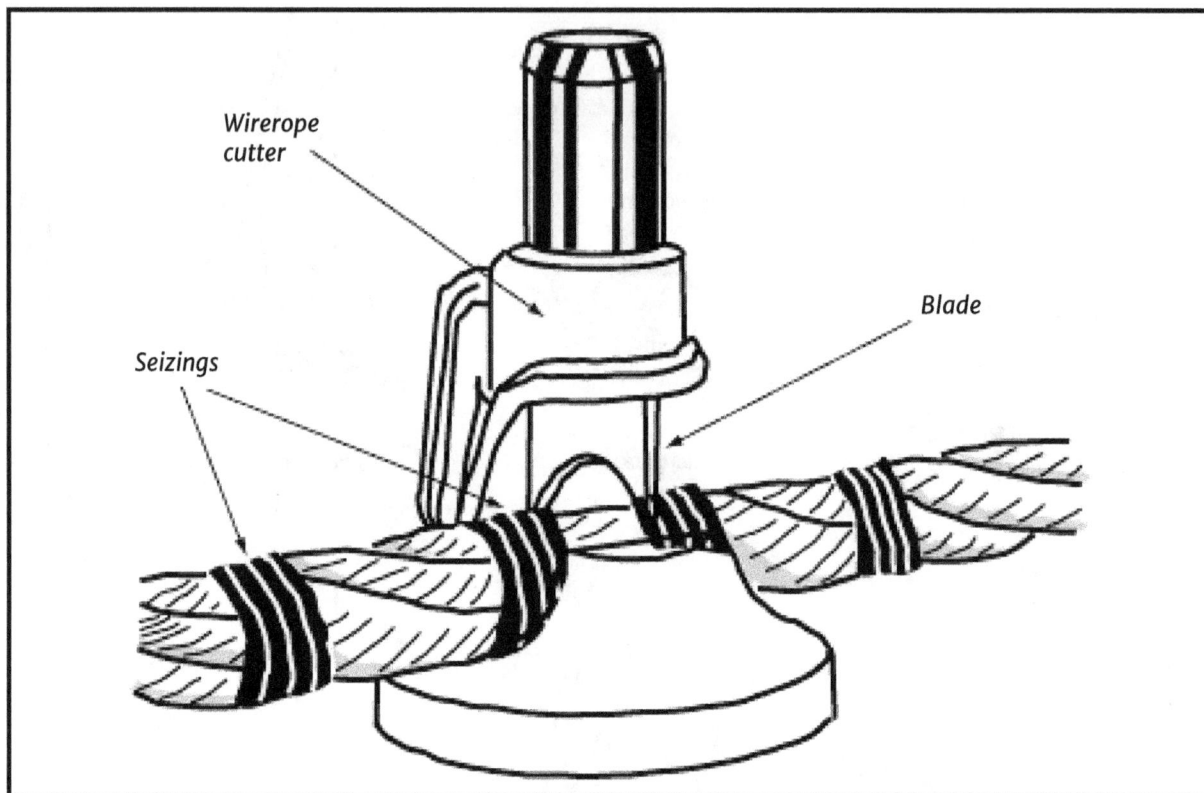

Figure 1-11. Wire-rope cutter

the two central seizings. Push the blade down against the wire rope and strike the top of the blade sharply with a sledgehammer several times. Use the bolt clipper son wire rope of fairly small diameter; however, use an oxyacetylene torch on wire rope of any diameter. The hacksaw and cold chisel are slower methods of cutting.

DRUMS AND SHEAVES

The size and location of the sheaves and drums about which wire rope operates and the speed with which the rope passes over the sheaves have a definite effect on the rope's strength and service life.

Size

Each time wire rope is bent, the individual strands must move with respect to each other in addition to bending. Keep this bending and moving of wires to a minimum to reduce wear. If the sheave or drum diameter is sufficiently large, the loss of strength due to bending wire rope around it will be about 5 or 6 percent. In all cases, keep the speed of the rope over the sheaves or drum as slow as is consistent with efficient work to decrease wear on the rope. It is impossible to give an absolute minimum size for each sheave or drum, since a number of factors enter into this decision. However, *Table 1-4, page 1-16*, shows the minimum recommended sheave and drum diameters for several wire-rope sizes. The sheave diameter always should be as large as possible and, except for very flexible rope, never less than 20 times the wire-rope diameter. This figure has been adopted widely.

Table 1-4. Minimum tread diameter of drums and sheaves

Rope Diameter (inches)	Minimum Tread Diameter for Given Rope Construction* (inches)			
	6 x 7	6 x 19	6 x 37	8 x 19
1/4	10 1/2	8 1/2		6 1/2
3/8	15 3/4	12 3/4	6 3/4	9 3/4
1/2	21	17	9	13
5/8	26 1/4	21 1/4	11 1/4	16 1/4
3/4	31 1/2	25 1/2	13 1/2	19 1/2
7/8	36 3/4	29 3/4	15 3/4	22 3/4
1	42	34	18	26
1 1/8	47 1/4	38 1/4	20 1/4	29 1/4
1 1/4	52 1/2	42 1/2	22 1/2	32 1/2
1 1/2	63	51	27	39

*Rope construction is strands and wires per strand.

Location

You should reeve the drums, sheaves, and blocks used with wire rope and place them in a manner to avoid reverse bends whenever possible (see *Figure 1-12*). A reverse bend occurs when rope bends in one direction around one block, drum, or sheave and bends in the opposite direction around the next. This causes the individual wires and strands to do an unnecessary amount of shifting, which increases wear. Where you must use a reverse bend, the block or drum causing the reversal should be of larger diameter than ordinarily used. Space the bends as far apart as possible so there will be more time allowed between the bending motions.

Winding

Do not overlap wire-rope turns when winding them on the drum of a winch; wrap them in smooth layers. Overlapping results in binding, causing snatches on the line when the rope is unwound. To produce smooth layers, start the rope against one flange of the drum and keep tension on the line while winding. Start the rope against the right or left flange as necessary to match the direction of winding, so that when it is rewound on the drum, the rope will curve in the same manner as when it left the reel (see *Figure 1-13*). A convenient method for determining the proper flange of the drum for starting the rope is known as the hand rule (see *Figure 1-14, page 1-18*). The extended index finger in this figure points at the on-winding rope. The turns of the rope are wound on the drum close together to prevent the possibility of crushing and abrasion of the rope while it is winding and to prevent binding or snatching when it is unwound. If necessary, use a wood stick to force the

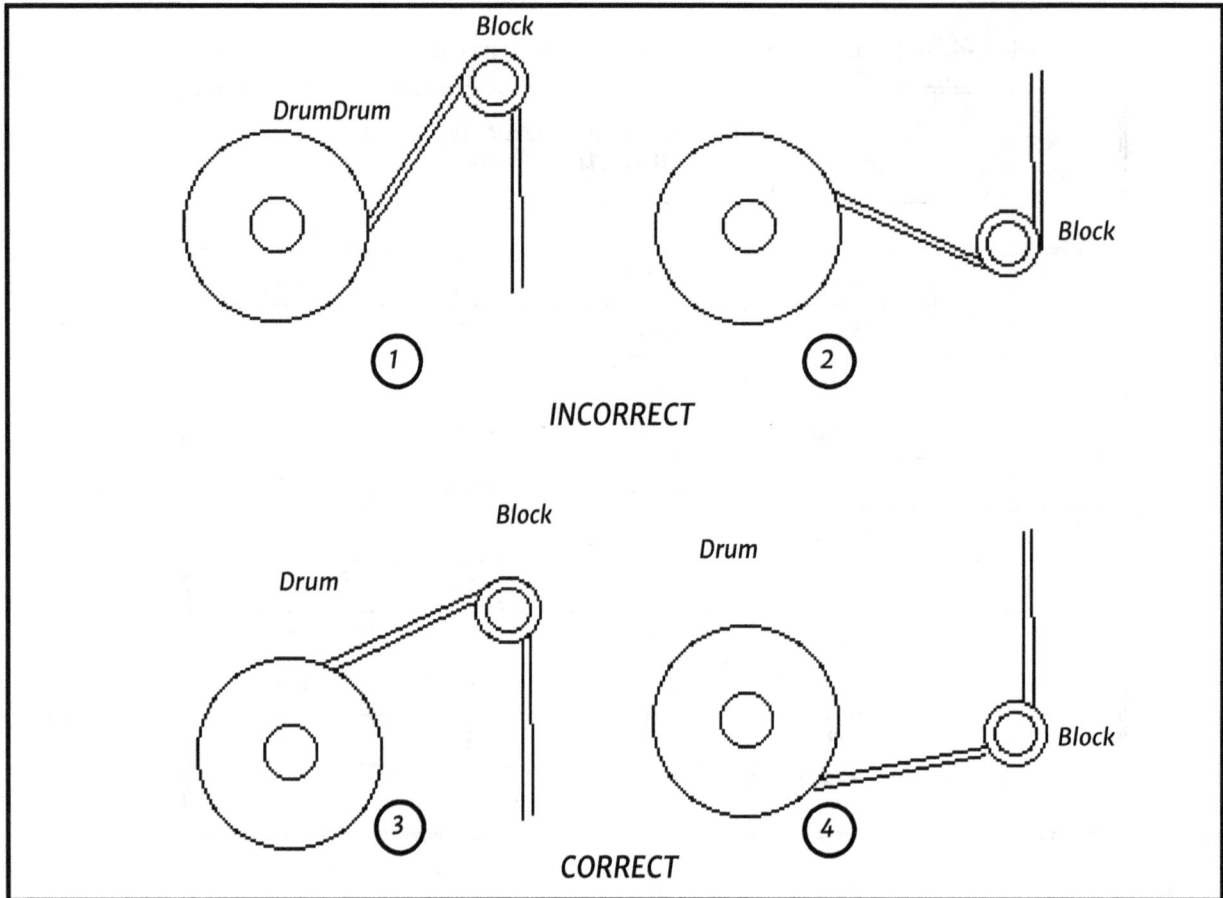

Figure 1-12. Avoiding reverse bends in wire rope

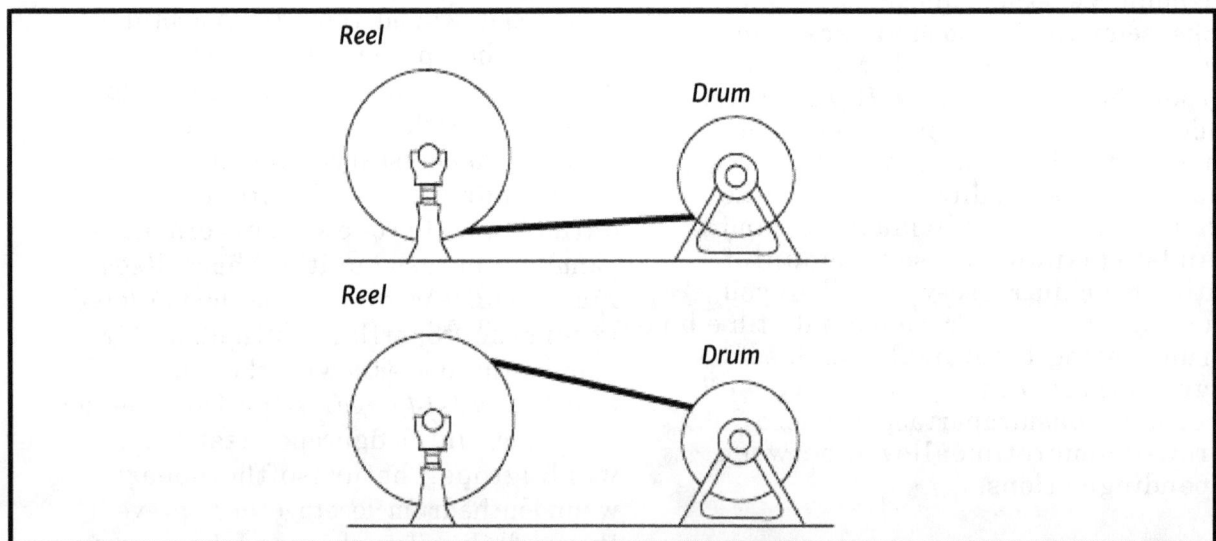

Figure 1-13. Spooling wire rope from reel to drum

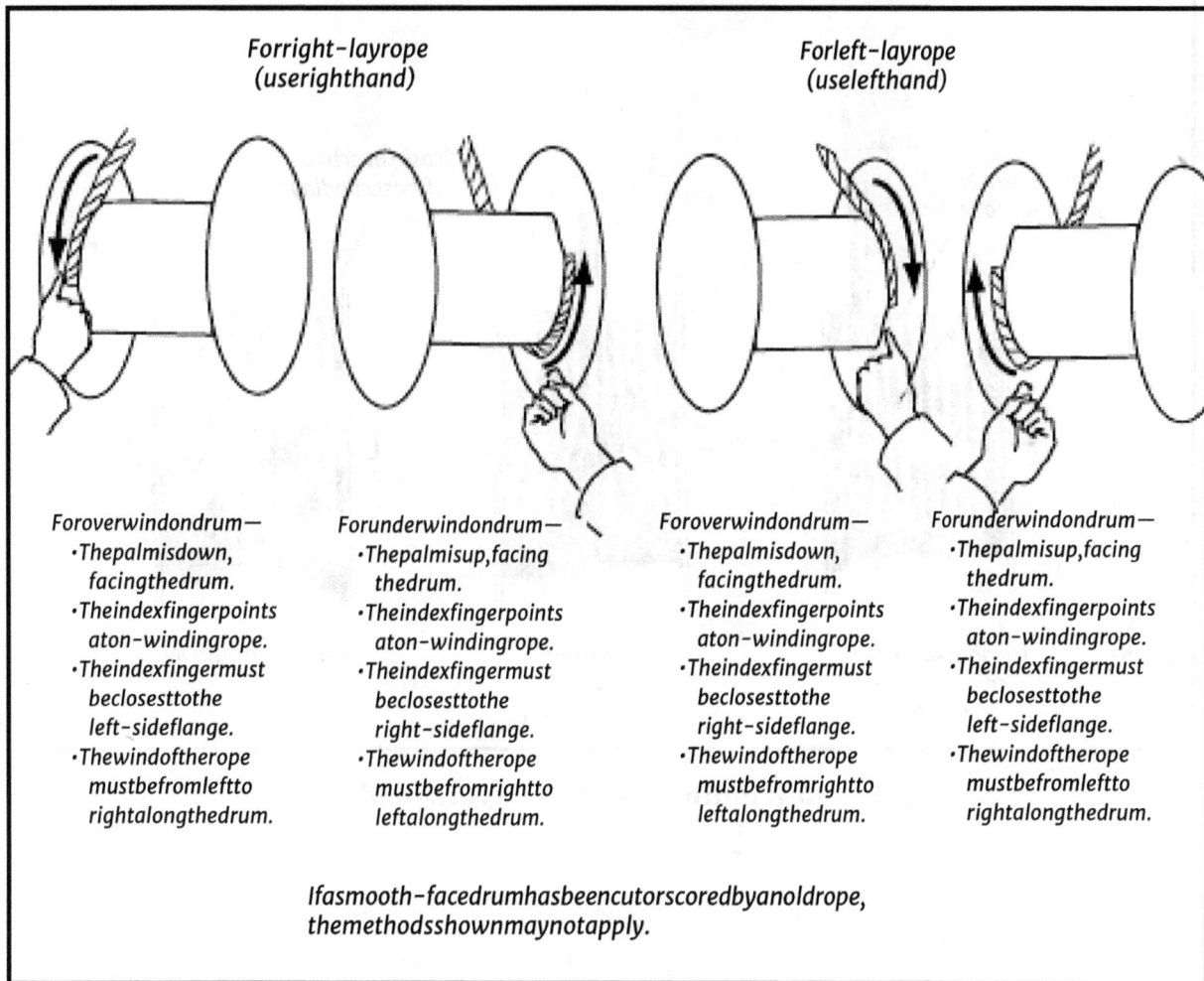

Figure 1-14. Determining starting flange of wire rope

turns closer together. Striking the wire with a hammer or other metal object damages the individual wires in the rope. If possible, wind only a single layer of wire rope on the drum. Where it is necessary to wind additional layers, wind them so as to eliminate the binding. Wind the second layer of turns over the first layer by placing the wire in the grooves formed by the first layer; however, cross each turn of the rope in the second layer over two turns of the first layer (see *Figure 1-15*). Wind the third layer in the grooves of the second layer; however, each turn of the rope will cross over two turns of the second layer.

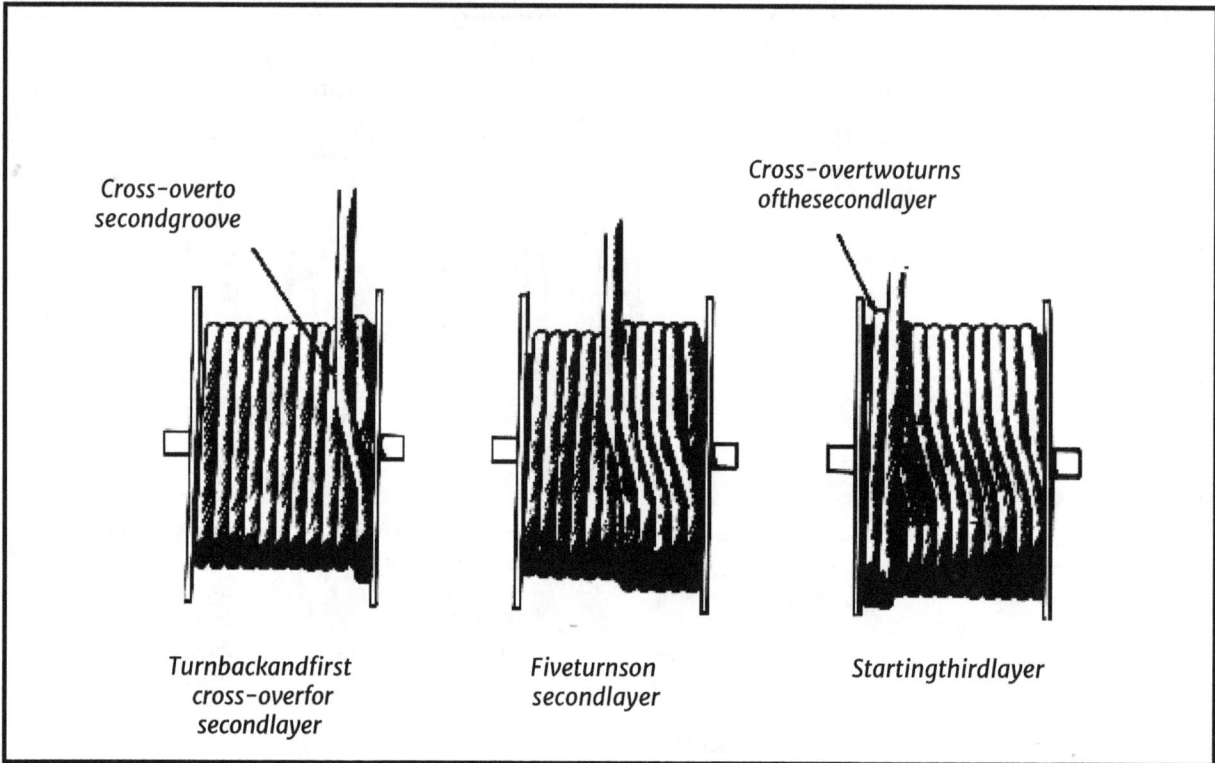

Figure1-15.Windingwire-ropelayersonadrum

INSPECTION OF WIRE ROPE

Inspect wire rope frequently. Replace frayed, kinked, worn, or corroded rope. The frequency of inspection is determined by the amount of use. A rope that is used 1 or 2 hours a week requires less frequent inspection than one that is used 24 hours a day.

PROCEDURES

Carefully inspect the weak points in rope and the points where the greatest stress occurs. Worn spots will show up as shiny flattened spots on the wires.

Inspect broken wires to determine whether it is a single broken wire or several wires. Rope is unsafe if—

- Individual wires are broken next to one another, causing unequal load distribution at this point.

- Replace the wire rope when 2.5 percent of the total rope wires are broken in the length of one lay, which is the length along the rope that a strand makes one complete spiral around the rope core. See Figure 1-16.

- Replace the wire rope when 1.25 percent of the total rope wires are broken in one strand in one lay.

- Replace wire rope with 200 or more wires (6x37 class) when the surface wires show flat wear spots equal in width to 80 percent of the diameter of the wires. On wire rope with larger and fewer total wires (6x7, 7x7, 7x19), replace it when the flat wear spot width is 50 percent of the wire diameter.

- Replace the wire if it is kinked or if there is evidence of a popped core or broken wire strands protruding from the core strand. See Figure 1-17.

- Replace the wire rope if there is evidence of an electrical arc strike (or other thermal damage) or crushing damage.

- Replace the wire rope if there is evidence of "birdcage" damage due to shock unloading. See Figure 1-17.

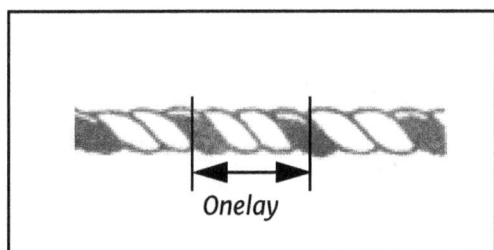

Onelay

Figure 1-16. Lay length

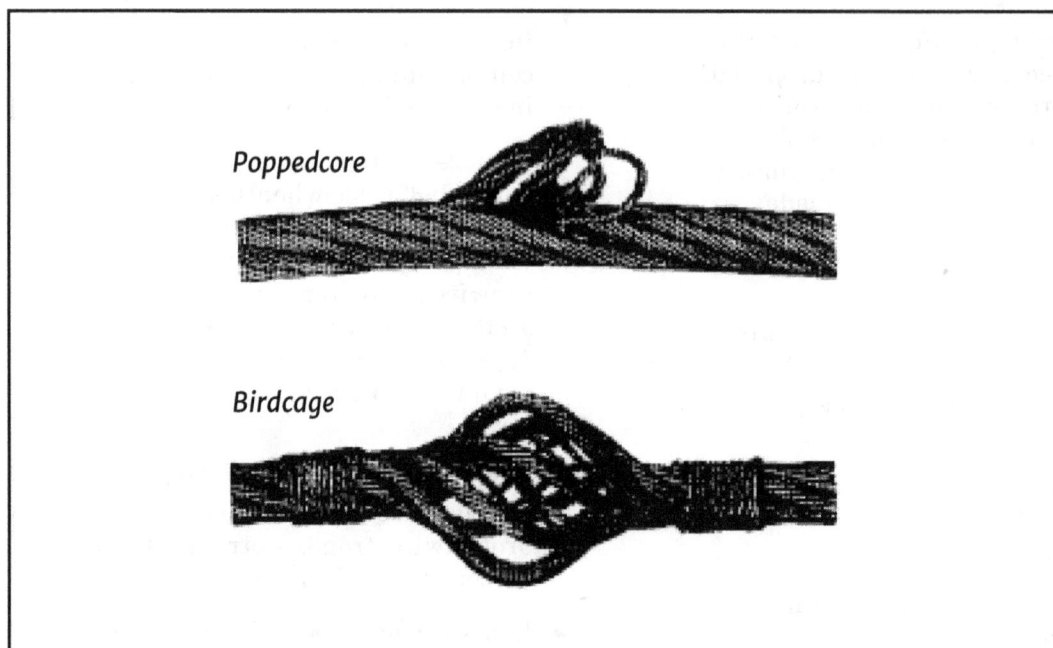

Figure1-17.Unserviceablewirerope

CAUSESOFFAILURE

Wireropefailureiscommonlycausedby—

- Sizing,constructing,orgradingit incorrectly.
- Allowingittodragoverobstacles.
- Lubricatingitimproperly.
- Operatingitoverdrumsandsheaves ofinadequatesize.
- Overwindingorcrosswindingiton drums.
- Operatingitoverdrumsandsheaves thatareoutofalignment.
- Permittingittojumpsheaves.
- Subjectingittomoistureoracid fumes.
- Permittingittountwist.
- Kinking.

CHAPTER 2

Knots, Splices, Attachments, and Ladders

Section I. Knots, Hitches, and Lashings

A study of the terminology pictured in *Figure 2-1* and the definitions in *Table 2-1, page 2-2,* will aid in understanding the methods of knotting presented in this section.

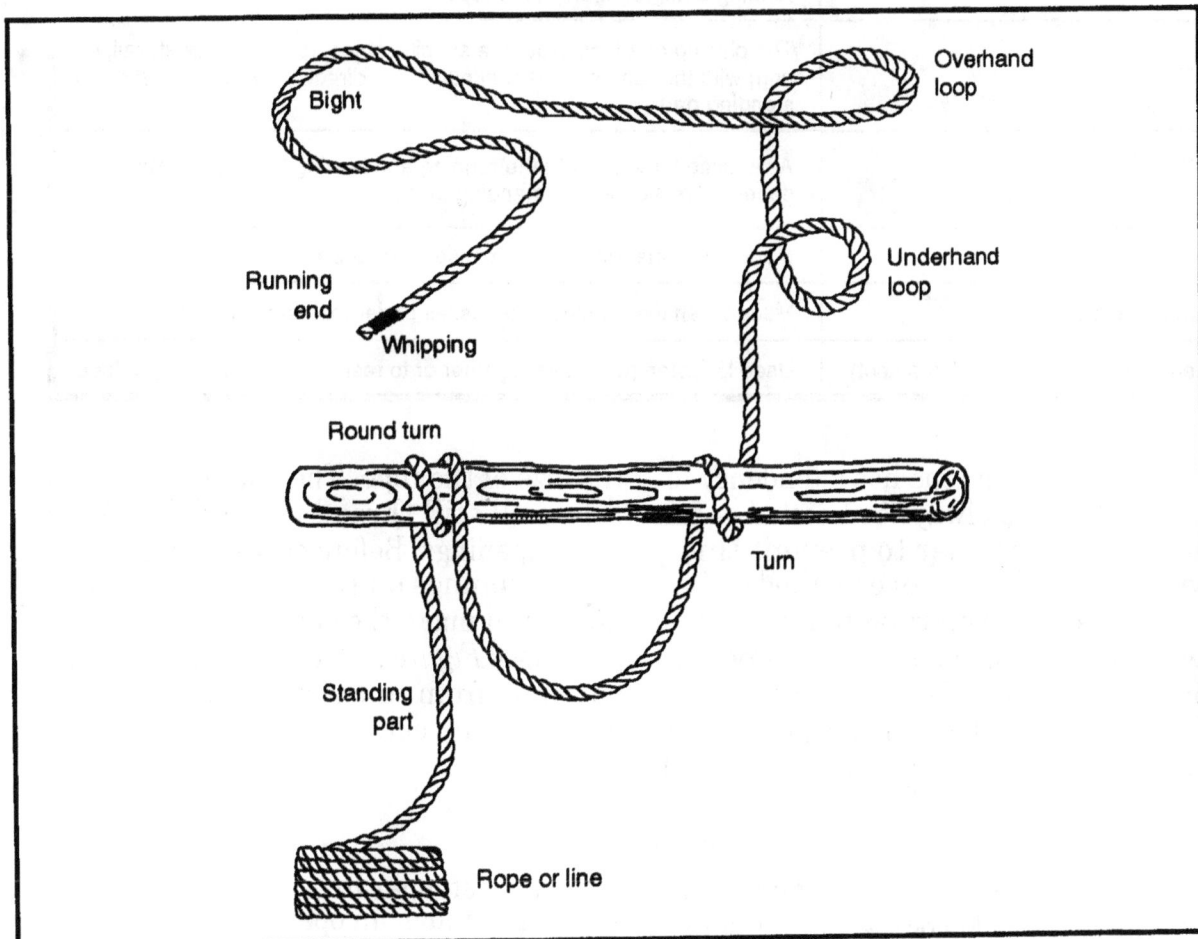

Figure 2-1. Elements of knots, bends, and hitches

Table 2-1. Knotting terminology and definitions

Term	Definition
Rope (often called a line)	A large, stout cord made of strands of fiber or wire that are twisted or braided together
Line (sometimes called a rope)	A thread, string, cord, or rope, especially a comparatively slender and strong cord (This manual will use the word rope rather than line in describing knots, hitches, rigging, and the like.)
Running end	The free or working end of a rope
Standing part	The rest of the rope, excluding the running end
Bight	A bend or U-shaped curve in a rope
Loop	Formed by crossing the running end over or under the standing part, forming a ring or circle in the rope
Turn	The placing of a loop around a specific object (such as a post, rail, or ring) with the running end continuing in a direction opposite to the standing part
Round turn	A modified turn, but with the running end leaving the circle in the same general direction as the standing part
Overhand turn or loop	Made when the running end passes over the standing part
Underhand turn or loop	Made when the running end passes under the standing part
Bend (in this manual, called a knot)	Used to fasten two ropes together or to fasten a rope to a ring or loop

The raw, cut end of a rope has a tendency to untwist and should always be knotted or fastened in some manner to prevent this untwisting. Whipping is one method of fastening the end of the rope to prevent untwisting (see *Figure 2-2*). A rope is whipped by wrapping the end tightly with a small cord. This method is particularly satisfactory because there is very little increase in the size of the rope. The whipped end of a rope will still thread through blocks or other openings. Before cutting a rope, place two whippings on the rope 1 or 2 inches apart and make the cut between the whippings (see *Figure 2-2*). This will prevent the cut ends from untwisting immediately after they are cut.

KNOTS

A knot is an interlacement of the parts of one or more flexible bodies, such as cordage rope, forming a lump. It is also any tie or fastening formed with a rope, including bends, hitches, and splices. A knot is often used as a stopper to prevent a rope from passing through an opening.

A good knot must be easy to tie, must hold without slipping, and must be easy to untie. The choice of the best knot, bend, or

Lay the bight along the rope.

End Start whipping here.

Lay the last round through the loop.

Pull loop to the center. Cut here.

(The loops are opened in this illustration to clarify the whipping procedure.)

Cut between whippings.

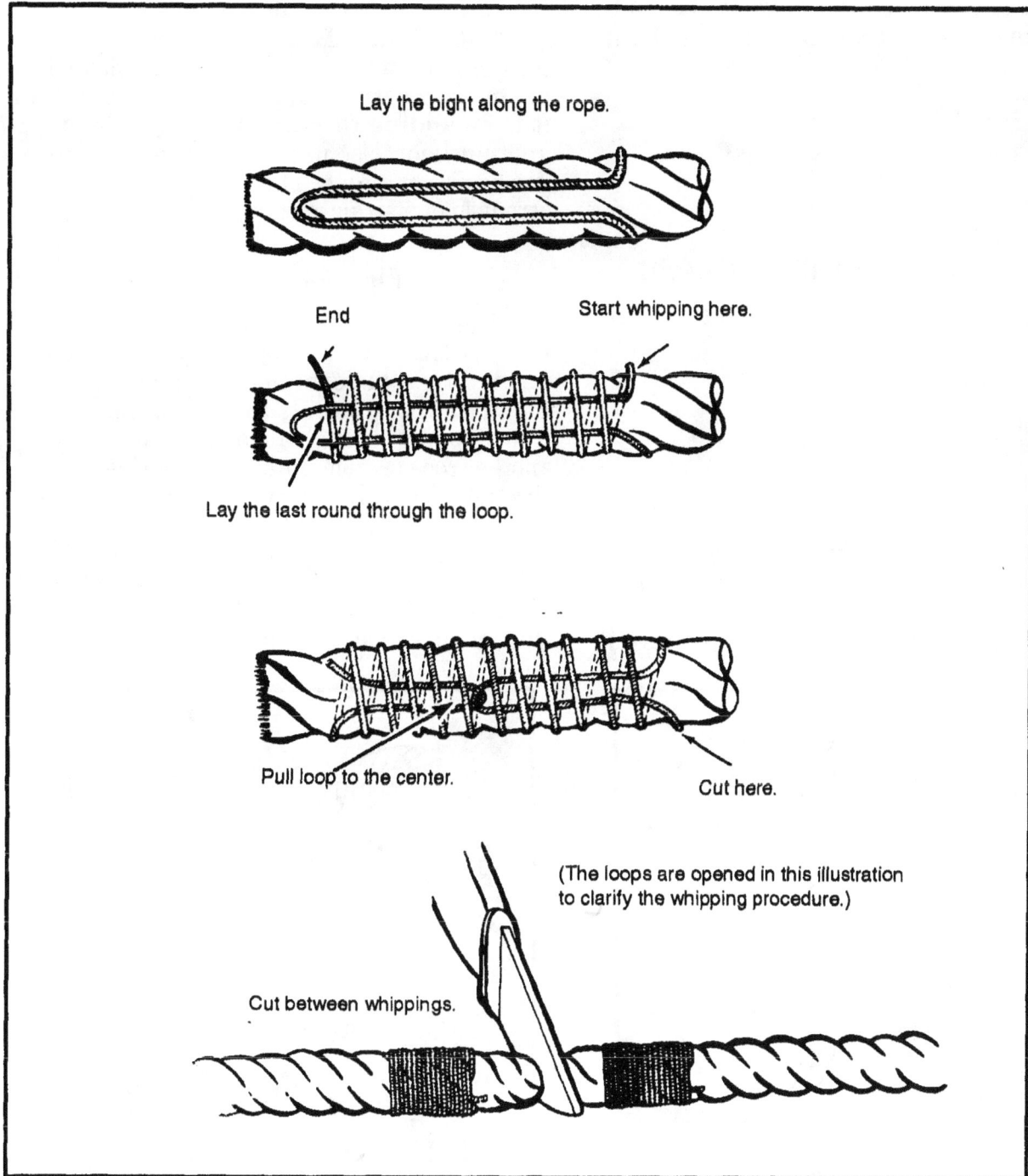

Figure 2-2. Whipping the end of a rope

hitch to use depends largely on the job it has to do. In general, knots can be classified into three groups. They are—

- Knots at the end of a rope.
- Knots for joining two ropes.
- Knots for making loops.

KNOTS AT THE END OF ROPE

Knots at the end of a rope fall into the following categories:

- Overhand knot.
- Figure-eight knot.
- Wall knot.

Overhand Knot

The overhand knot is the most commonly used and the simplest of all knots (see *Figure 2-3*). Use an overhand knot to prevent the end of a rope from untwisting, to form a knob at the end of a rope, or to serve as a part of another knot. When tied at the end or standing part of a rope, this knot prevents it from sliding through a block, hole, or another knot. Use it also to increase a person's grip on a rope. This knot reduces the strength of a straight rope by 55 percent.

Figure-Eight Knot

Use the figure-eight knot to form a larger knot at the end of a rope than would be formed by an overhand knot (see *Figure 2-4*). The knot prevents the end of the rope from slipping through a fastening or loop in another rope or from unreeving when reeved through blocks. It is easy to untie.

Wall Knot

Use the wall knot with crown to prevent the end of a rope from untwisting when an

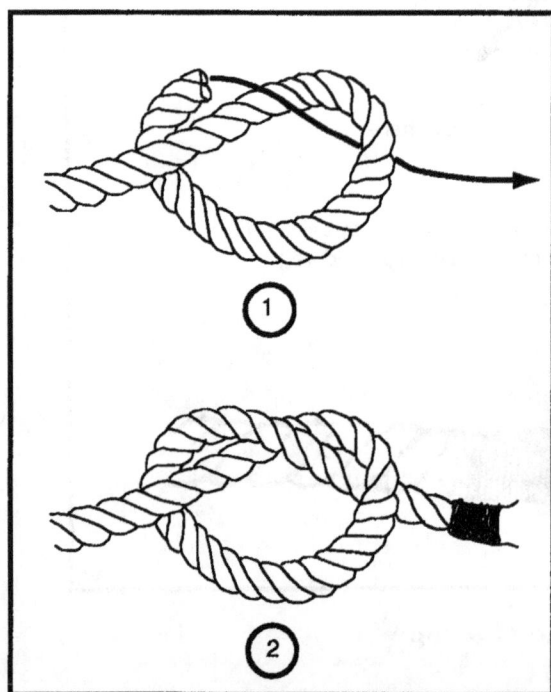

Figure 2-3. Overhand knot

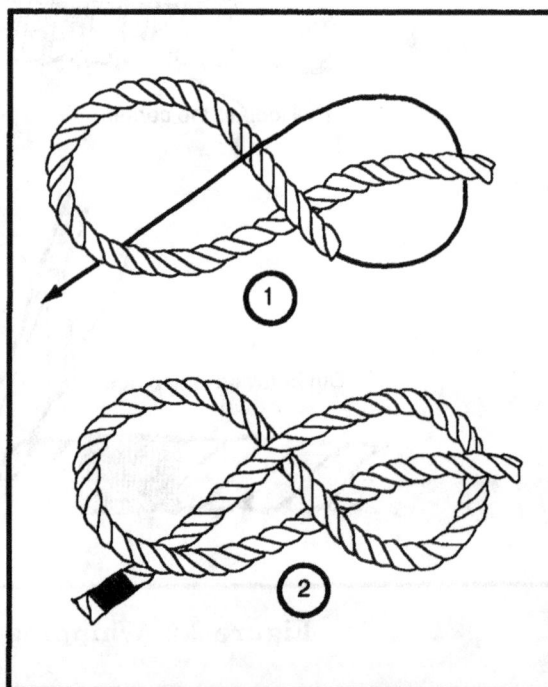

Figure 2-4. Figure-eight knot

enlarged end is not objectionable (see*Figure 2-5).* The wall knot also makes a desirable knot to prevent the end of the rope from slipping through small openings, as when using rope handles on boxes. Use either the crown or the wall knot separately to form semipermanent "stopper knots" tied with the end strands of a rope. The wall knot will prevent the rope from untwisting, but to make a neat round knob, crown it (see*Figure 2-6, page 2-6).*Notice that in the wall knot, the ends come up through the bights, causing the strands to lead forward. In a crown knot, the ends go down through the bights and point backward.

KNOTS FOR JOINING TWO ROPES

Knots for joining two ropes fall into the following categories:

- Square knot.
- Single sheet bend.
- Double sheet bend.
- Carrick bend.

Square Knot

Use the square knot to tie two ropes of equal size together so they will not slip (see

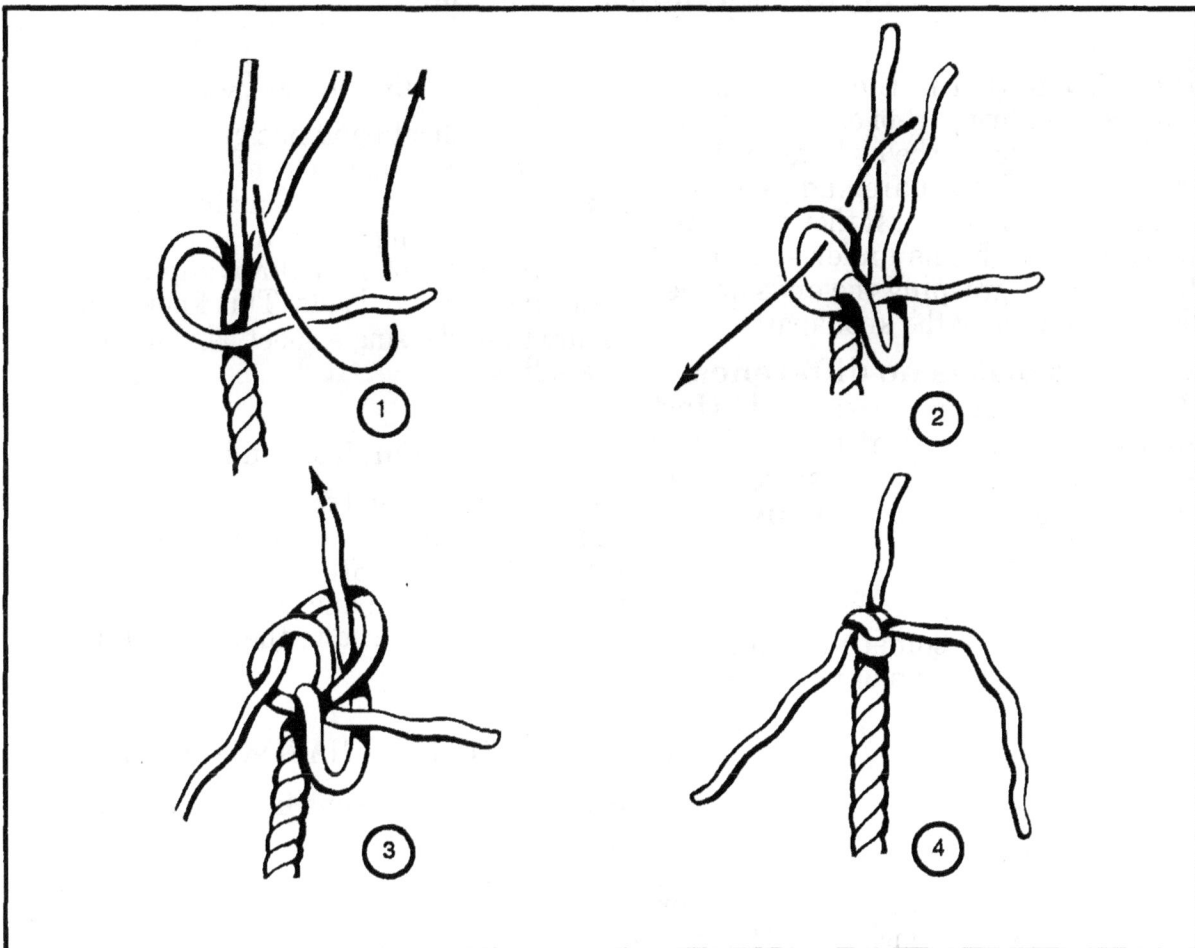

Figure 2-5. Wall knot

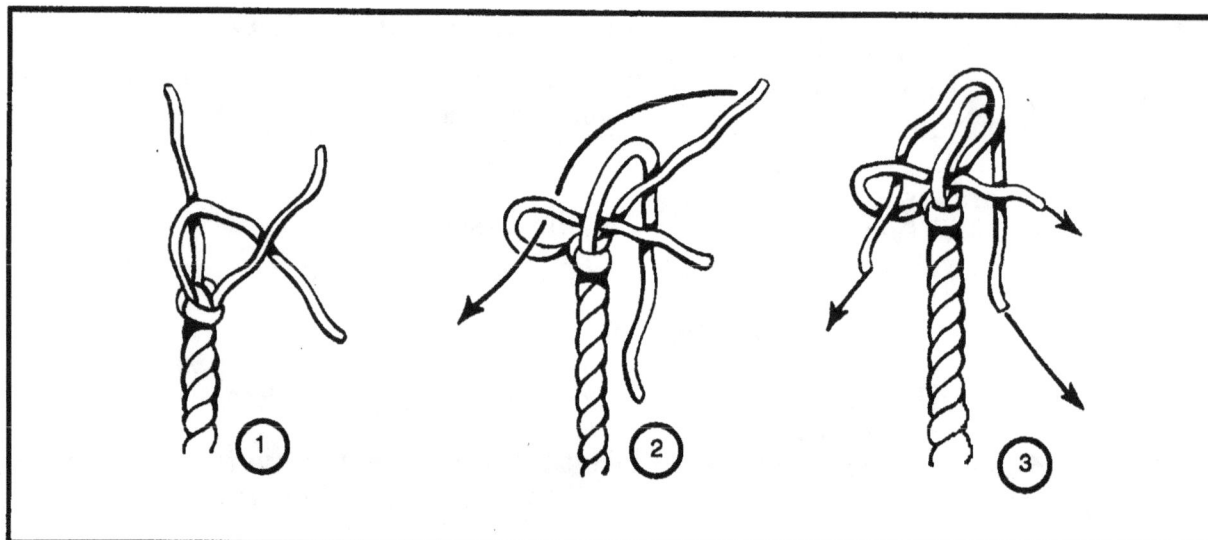

Figure 2-6. Crown on a wall knot

*Figure 2-7).*Note that in the square knot, the end and standing part of one rope come out on the same side of the bight formed by the other rope. The square knot will not hold if the ropes are wet or if they are of different sizes. It tightens under strain but can be untied by grasping the ends of the two bights and pulling the knot apart.

NOTE. It makes no difference whether the first crossing is tied left-over-right or right-over-left as long as the second crossing is tied opposite to the first crossing.

Single Sheet Bend

A single sheet bend, sometimes called a weaver's knot, has two major uses (see*Figure 2-8).*They are—

- Tying together two ropes of unequal size.

- Tying a rope to an eye.

This knot will draw tight but will loosen or slip when the lines are slackened. The single sheet bend is stronger and unties easier than the square knot.

Double Sheet Bend

The double sheet bend has greater holding power than the single sheet bend for joining ropes of equal or unequal diameter, joining wet ropes, or tying a rope to an eye (see*Figure 2-9, page 2-8,*)It will not slip or draw tight under heavy loads. This knot is more secure than the single sheet bend when used in a spliced eye.

Carrick Bend

Use the carrick bend for heavy loads and for joining large hawsers or heavy rope (see*Figure 2-10, page 2-8*)It will not draw tight under a heavy load and can be untied easily if the ends are seized to their own standing part.

KNOTS FOR MAKING LOOPS

Knots for making loops fall into the following categories:

- Bowline.

- Double bowline.

- Running bowline.

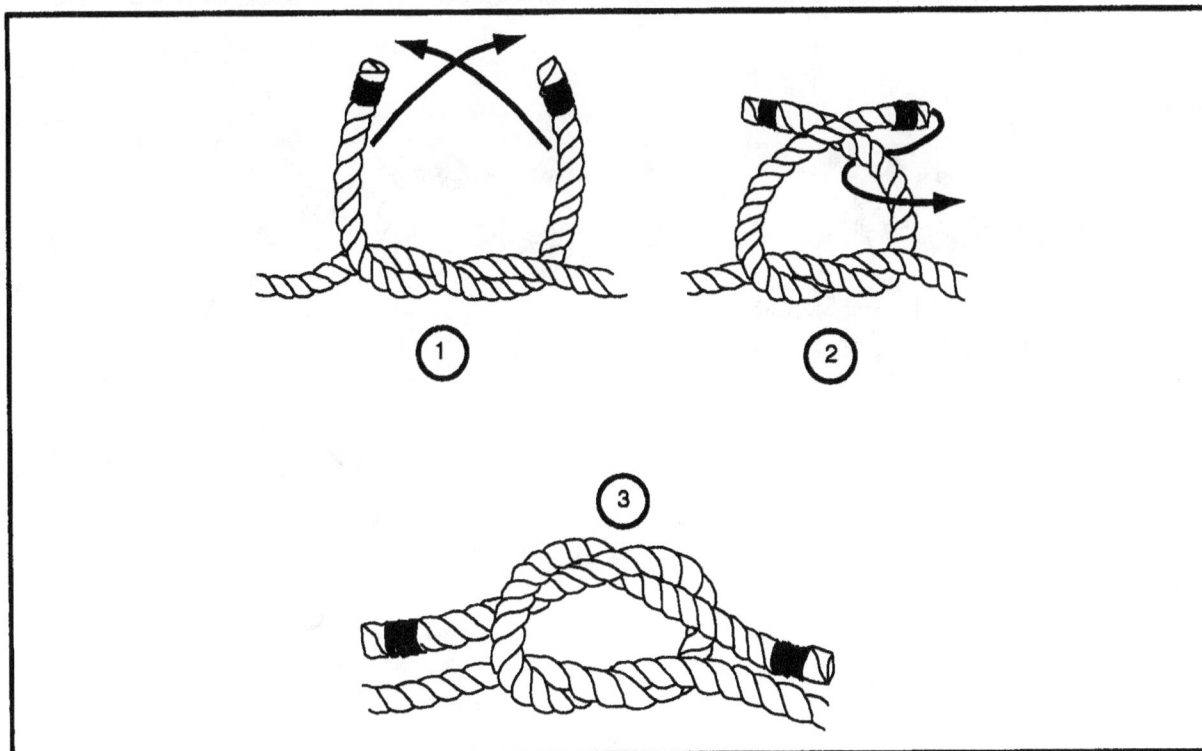

Figure 2-7. Square knot

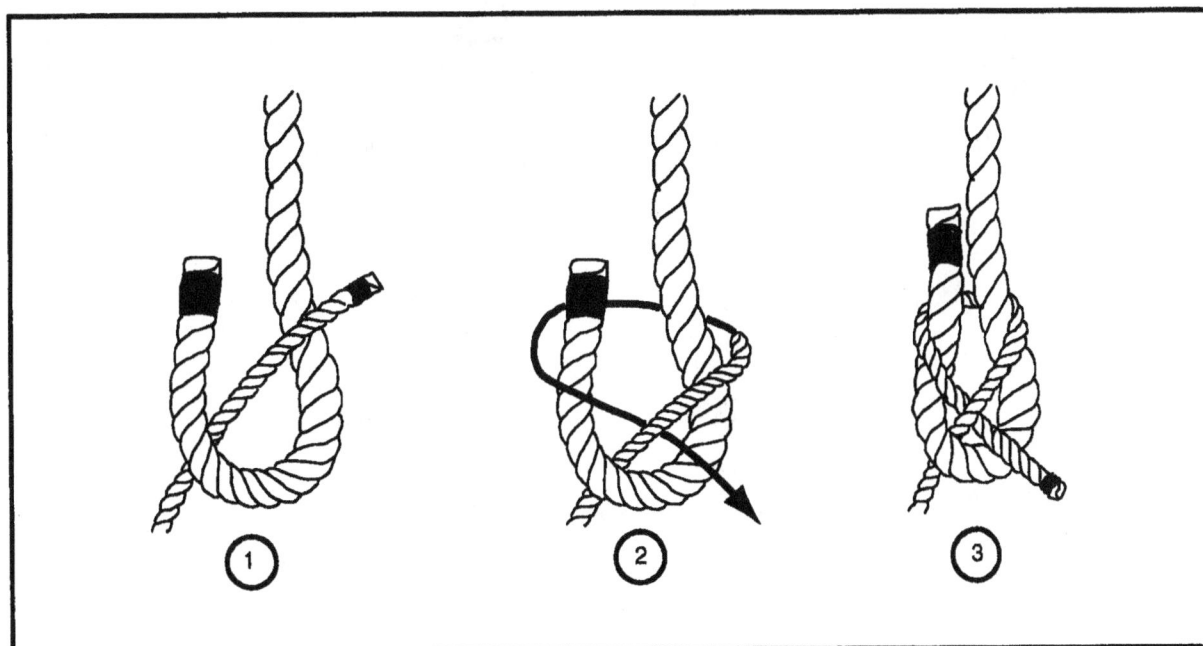

Figure 2-8. Single sheet bend

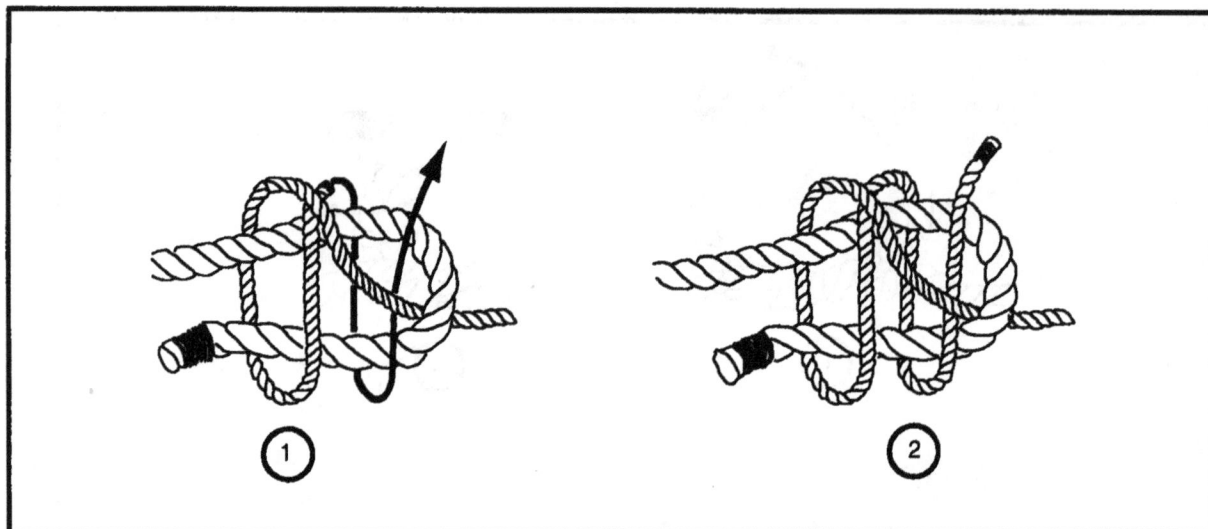

Figure 2-9. Double sheet bend

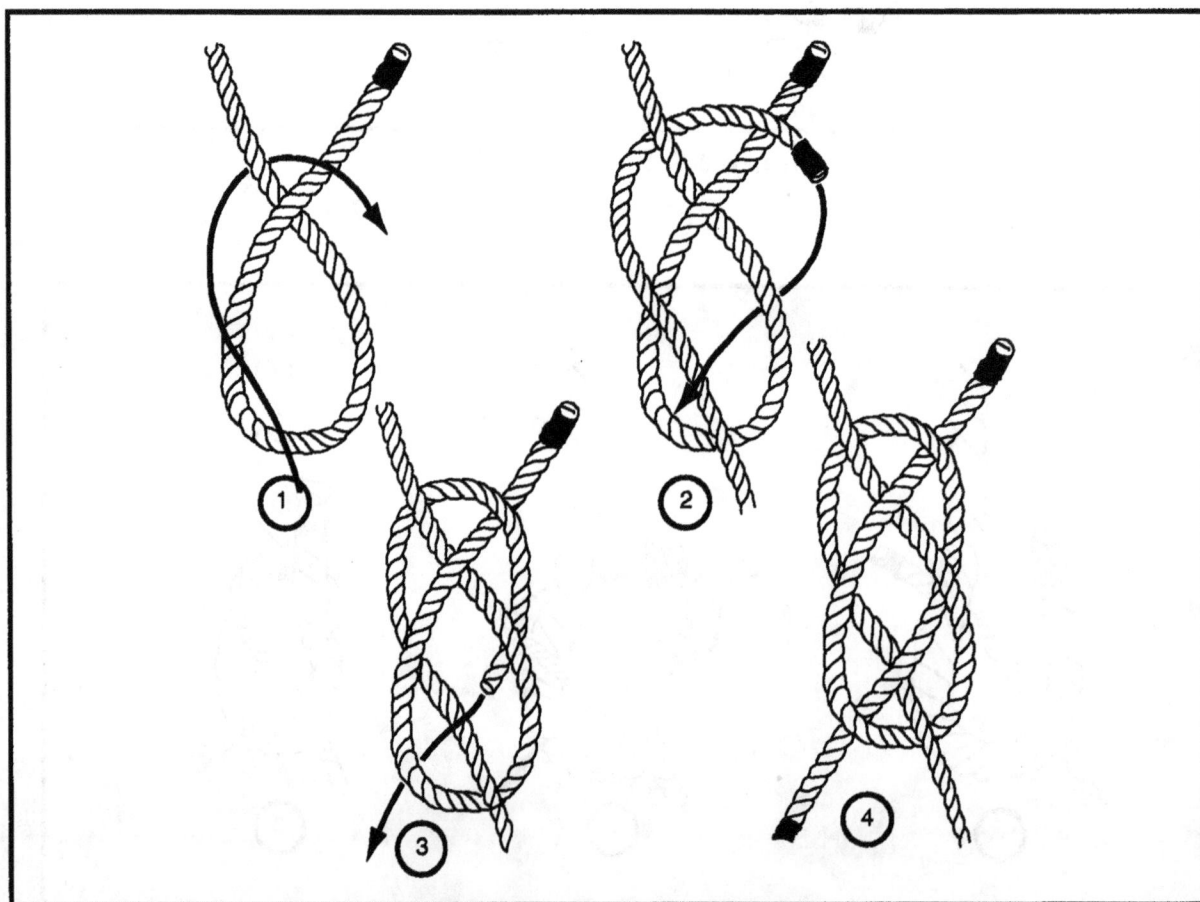

Figure 2-10. Carrick bend

- Bowline on a bight.
- Spanish bowline.
- French bowline.
- Speir knot.
- Cat's-paw.
- Figure eight with an extra turn.

Bowline

The bowline is one of the most common knots and has a variety of uses, one of which is the lowering of men and material (see *Figure 2-11*). It is the best knot for forming a single loop that will not tighten or slip under strain and can be untied easily if each running end is seized to its own standing part. The bowline forms a loop that may be of any length.

Double Bowline

The double bowline forms three nonslipping loops (see *Figure 2-12, page 2-10*). Use this knot to sling a man. As he sits in the slings, one loop supports his back and the remain-ing two loops support his legs. A notched board that passes through the two loops makes a comfortable seat known as a boatswain's chair. This chair is discussed in the scaffolding section of this manual (see *Chapter 6*).

Running Bowline

The running bowline forms a strong running loop (see *Figure 2-13, page 2-10*). It is a convenient form of running an eye. The running bowline provides a sling of the choker type at the end of a single line. Use it when tying a handline around an object at a point that you cannot safely reach, such as the end of a limb.

Bowline on a Bight

This knot forms two nonslipping loops (see *Figure 2-14, page 2-11*). You can use the bowline on a bight for the same purpose as a boatswain's chair. It does not leave both hands free, but its twin nonslipping loops form a comfortable seat. Use it when—

- You need more strength than a single bowline will give.

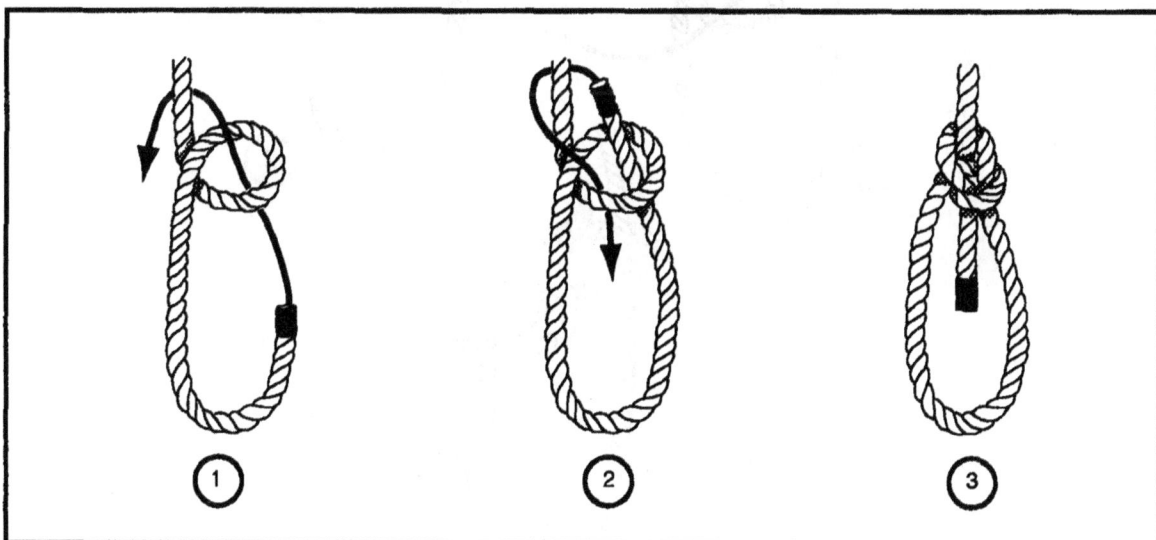

Figure 2-11. Bowline

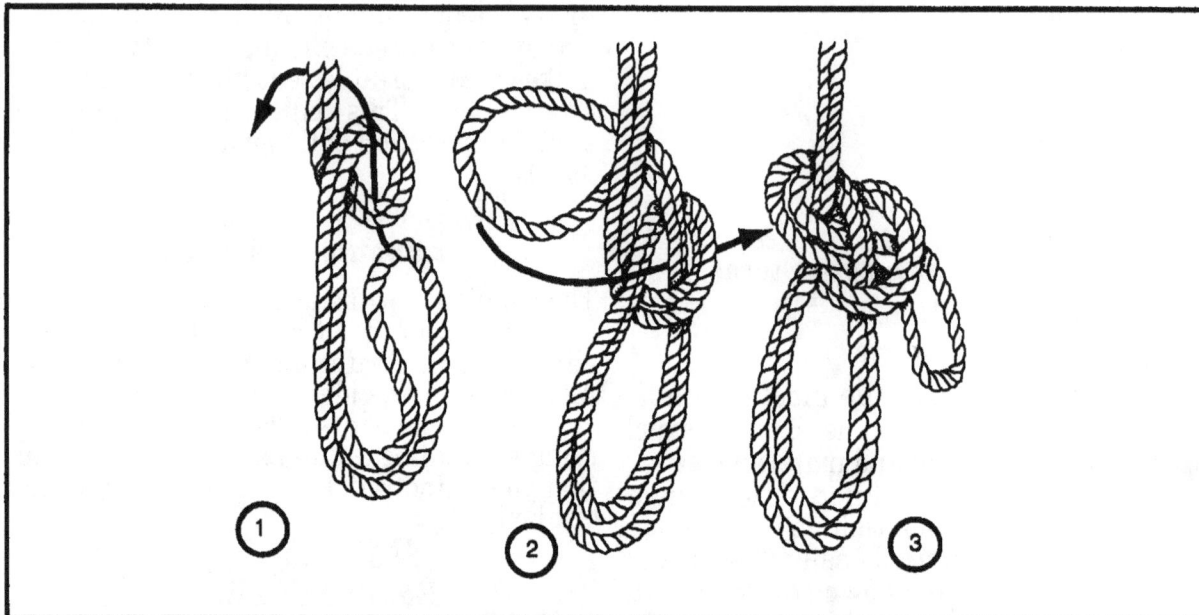

Figure 2-12. Double bowline

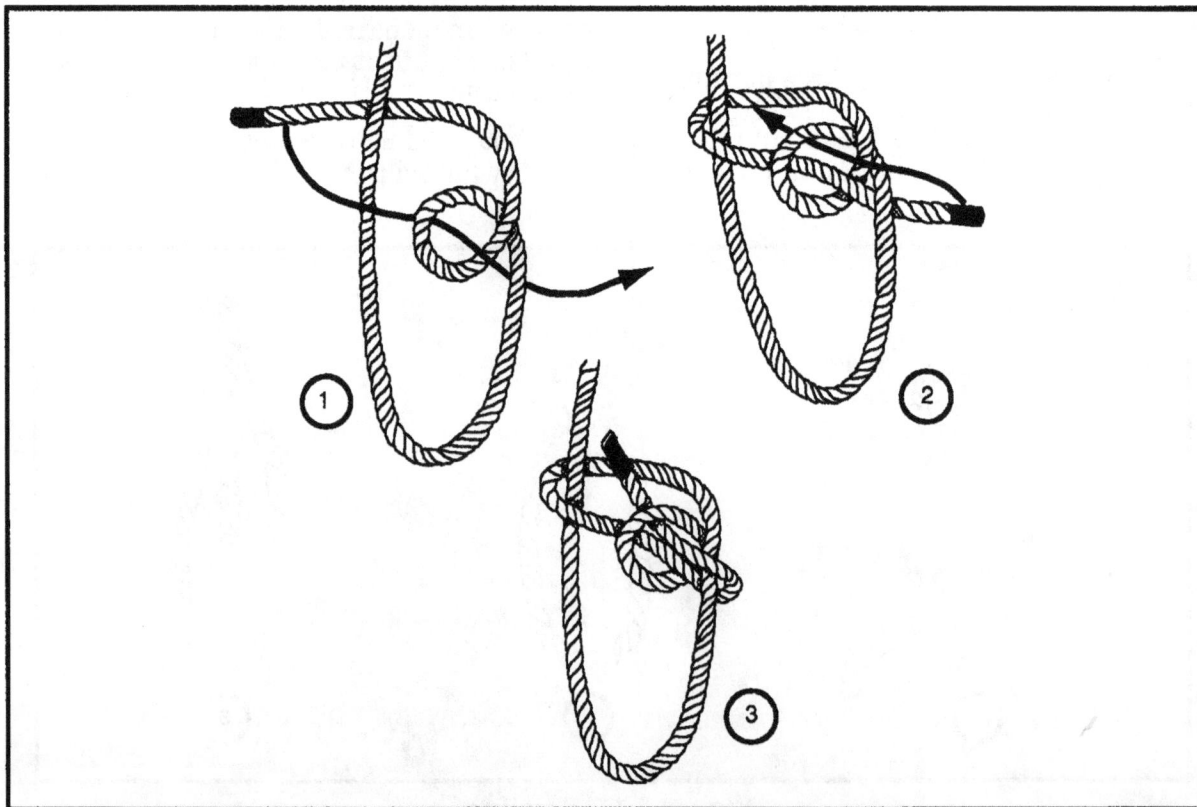

Figure 2-13. Running bowline

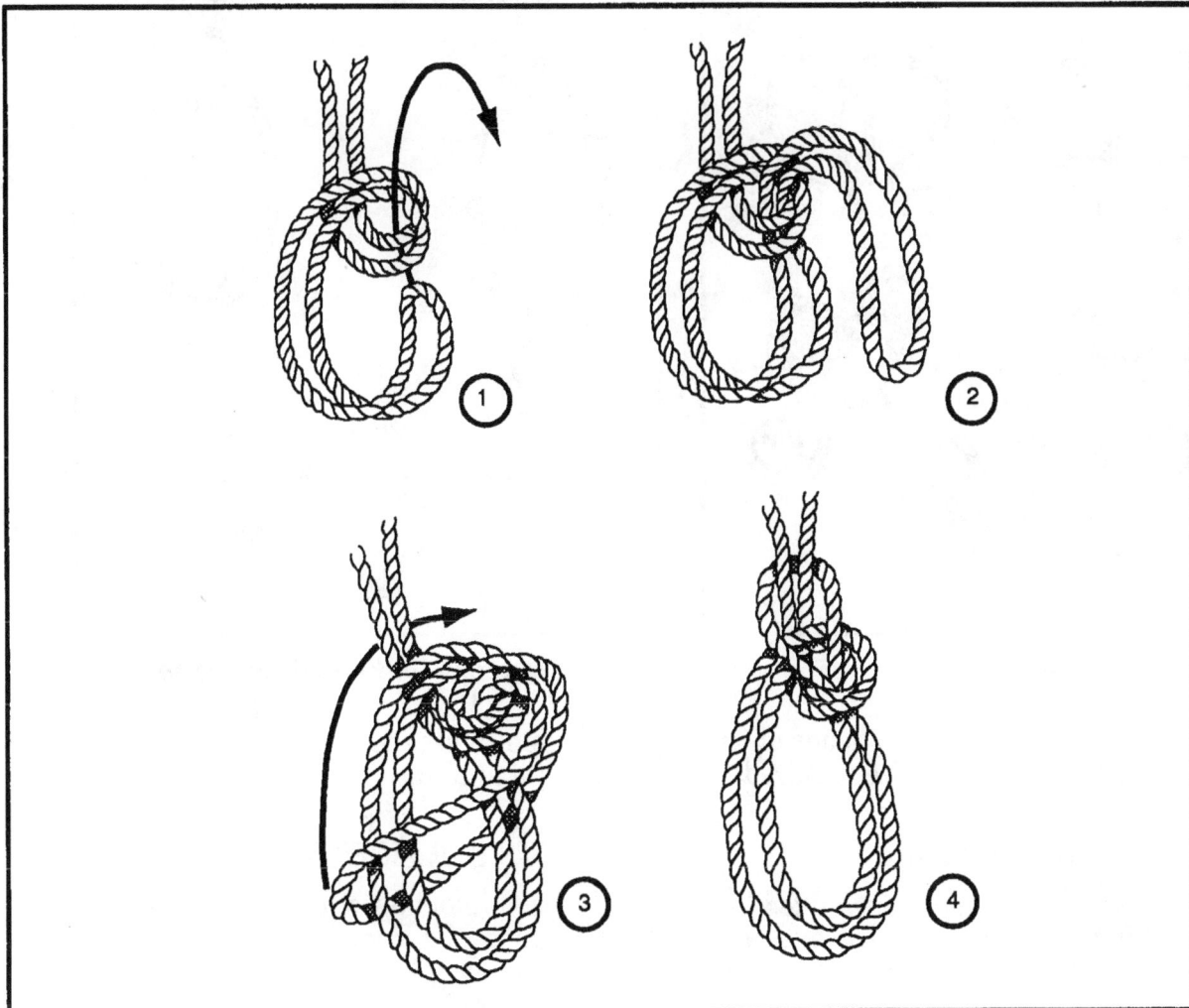

Figure 2-14. Bowline on a bight

- You need to form a loop at some point in a rope other than at the end.
- You do not have access to the end of a rope.

You can easily untie the bowline on a bight and tie it at the end of a rope by doubling the rope for a short section.

Spanish Bowline

You can tie a Spanish bowline at any point in a rope, either at a place where the line is double or at an end that has been doubled back (see *Figure 2-15, page 2-12*). Use the Spanish bowline in rescue work or to give a twofold grip for lifting a pipe or other round objects in a sling.

French Bowline

You can use the French bowline as a sling to lift injured men (see *Figure 2-16, page 2-12*). When used for this purpose, one loop is a seat and the other loop is put around the body under the arms. The injured man's weight keeps the two loops tight so that he cannot fall out. It is particularly useful as a sling for an unconscious man. Also, use the French bowline when working alone and you

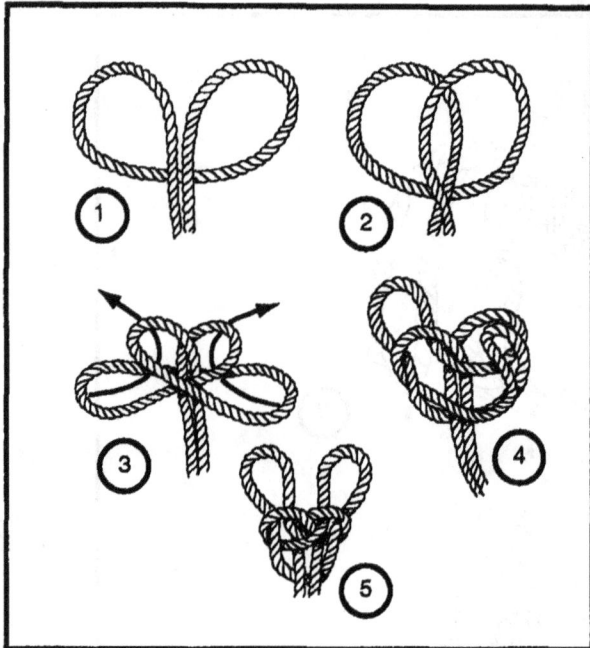

Figure 2-15. Spanish bowline

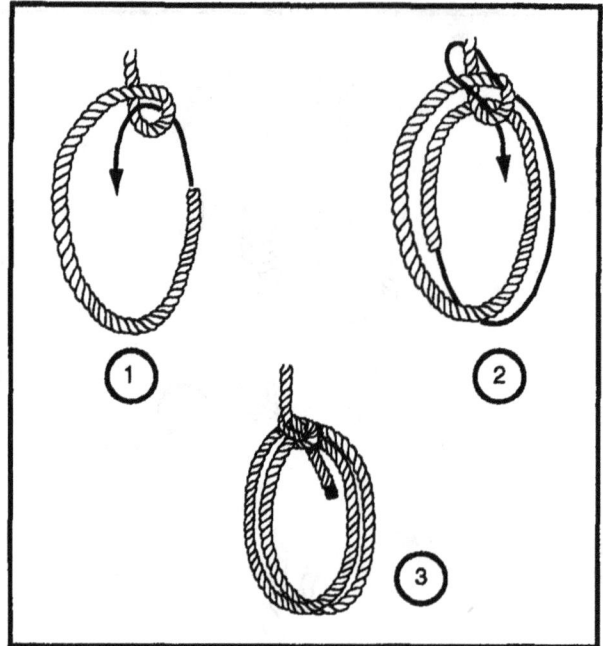

Figure 2-16. French bowline

need your hands free. The two loops of this knot can be adjusted to the size required.

Speir Knot

Use a speir knot when you need a fixed loop, a nonslip knot, and a quick release (see *Figure 2-17*). You can tie this knot quickly and release it by pulling on the running end.

Cat's-paw

Use a cat's-paw to fasten an endless sling to a hook, or make it at the end of a rope to fasten the rope to a hook (see *Figure 2-18*). You can tie or untie it easily. This knot, which is really a form of a hitch, is a more satisfactory way of attaching a rope to a hook than the blackwall hitch. It will not slip off and need not be kept taut to make it hold.

Figure Eight With an Extra Turn

Use a figure eight with an extra turn to tighten a rope (see *Figure 2-19, page 2-14*). This knot is especially suitable for tightening

a one-rope bridge across a small stream. You can tie and untie it easily.

KNOTS FOR TIGHTENING A ROPE

The types of knots used for tightening a rope are the butterfly knot and the baker bowline.

Butterfly Knot

Use the butterfly knot is to pull taut a high line, handline, tread rope for foot bridges, or similar installations (see *Figure 2-20, page 2-14*). Using this knot provides the capability to tighten a fixed rope when mechanical means are not available. (You can also use the harness hitch for this purpose [see *Figure 2-32, page 2-22*]. The butterfly knot will not jam if a stick is placed between the two upper loops.

Baker Bowline

You can use the baker bowline for the same purpose as the butterfly knot and for lashing cargo (see *Figure 2-21, pages 2-15 and 2-16*).

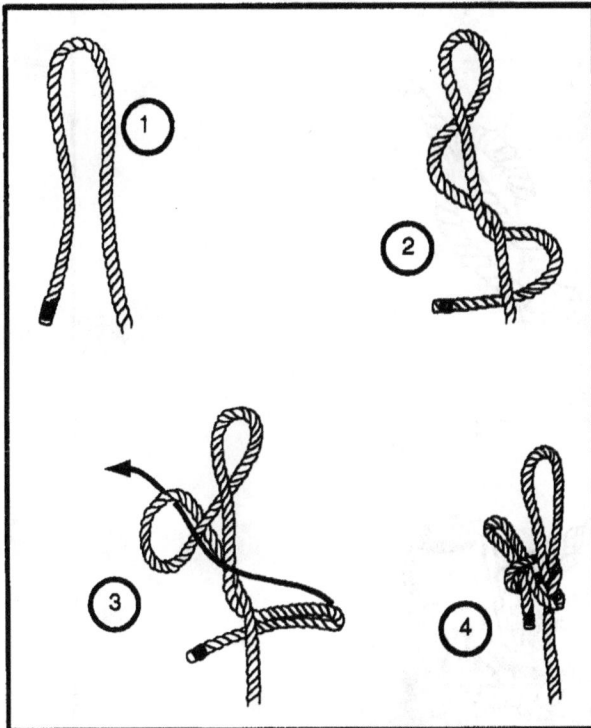

Figure 2-17. Speir knot

When used to lash cargo, secure one end with two half hitches, pass the rope over the cargo and tie a baker bowline, then secure the lashing with a slippery half hitch. To release the rope, simply pull on the running end. Advantages of the baker bowline are that it can be—

- Tied easily.

- Adjusted without losing control.

- Released quickly.

KNOTS FOR WIRE ROPE

Under special circumstances, when wire-rope fittings are not available and it is necessary to fasten wire rope by some other manner, you can use certain knots. In all knots made with wire rope, fasten the running end of the rope to the standing part after tying the knot. When wire-rope clips are available, use them to fasten the running end. If clips are not available, use

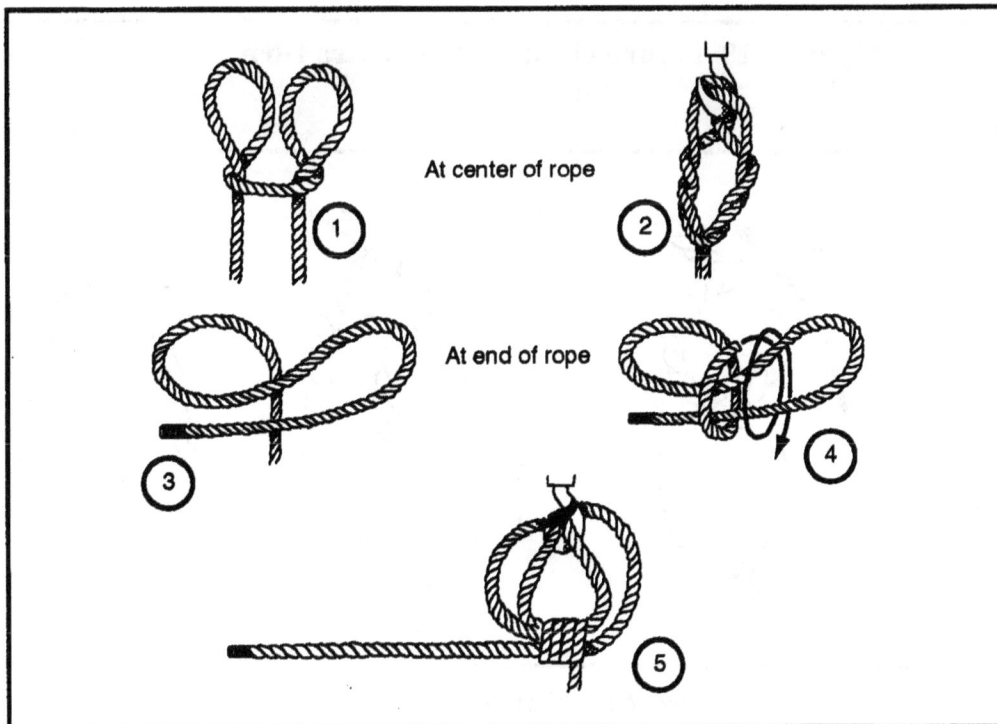

Figure 2-18. Cats-paw

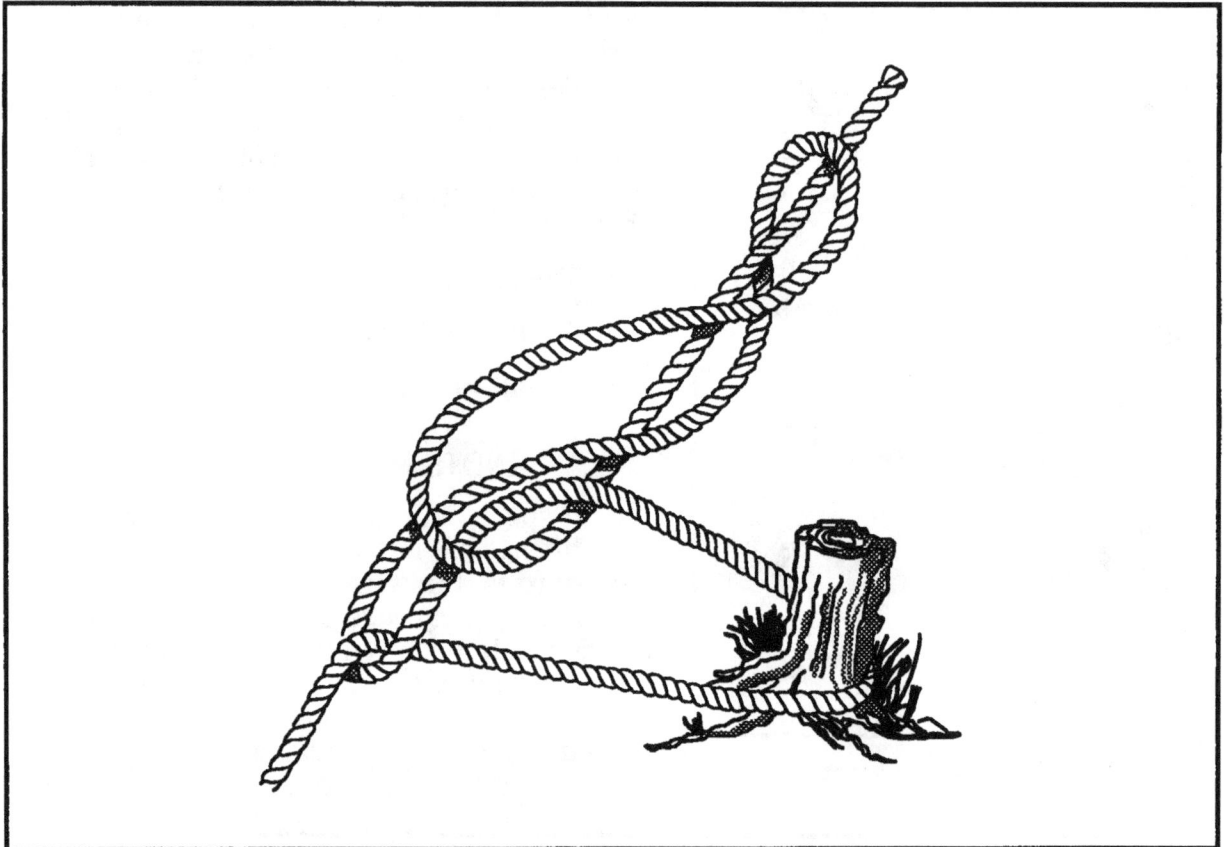

Figure 2-19. Figure eight with an extra turn

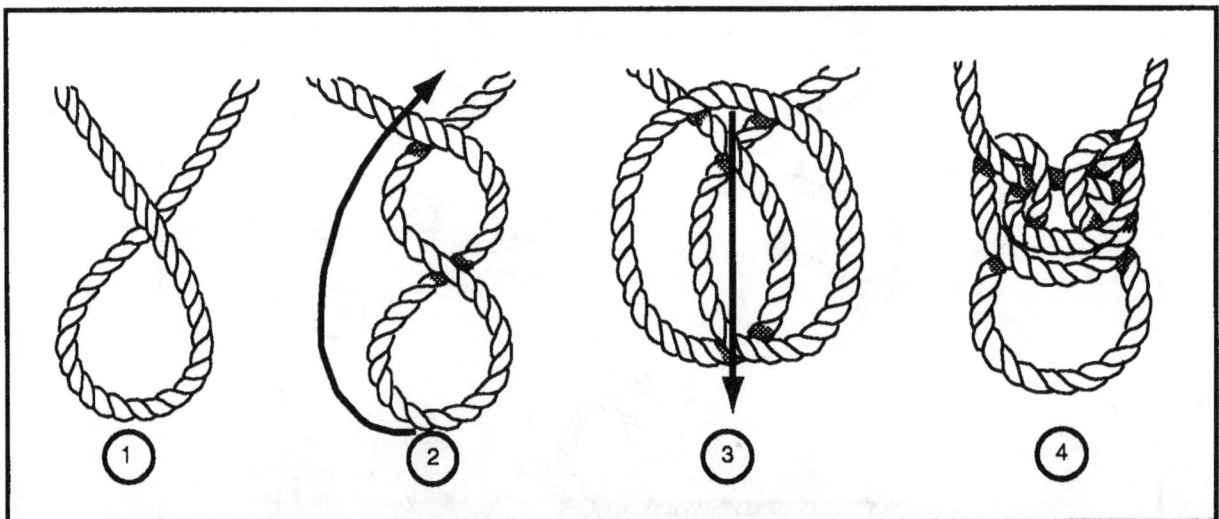

Figure 2-20. Butterfly knot

Form a loop.

Pass the running end under the object and through the tie-down ring.

Use a round turn and two half hitches.

①

②

Draw the bight through the loop.

Form a bight under the loop.

③

④

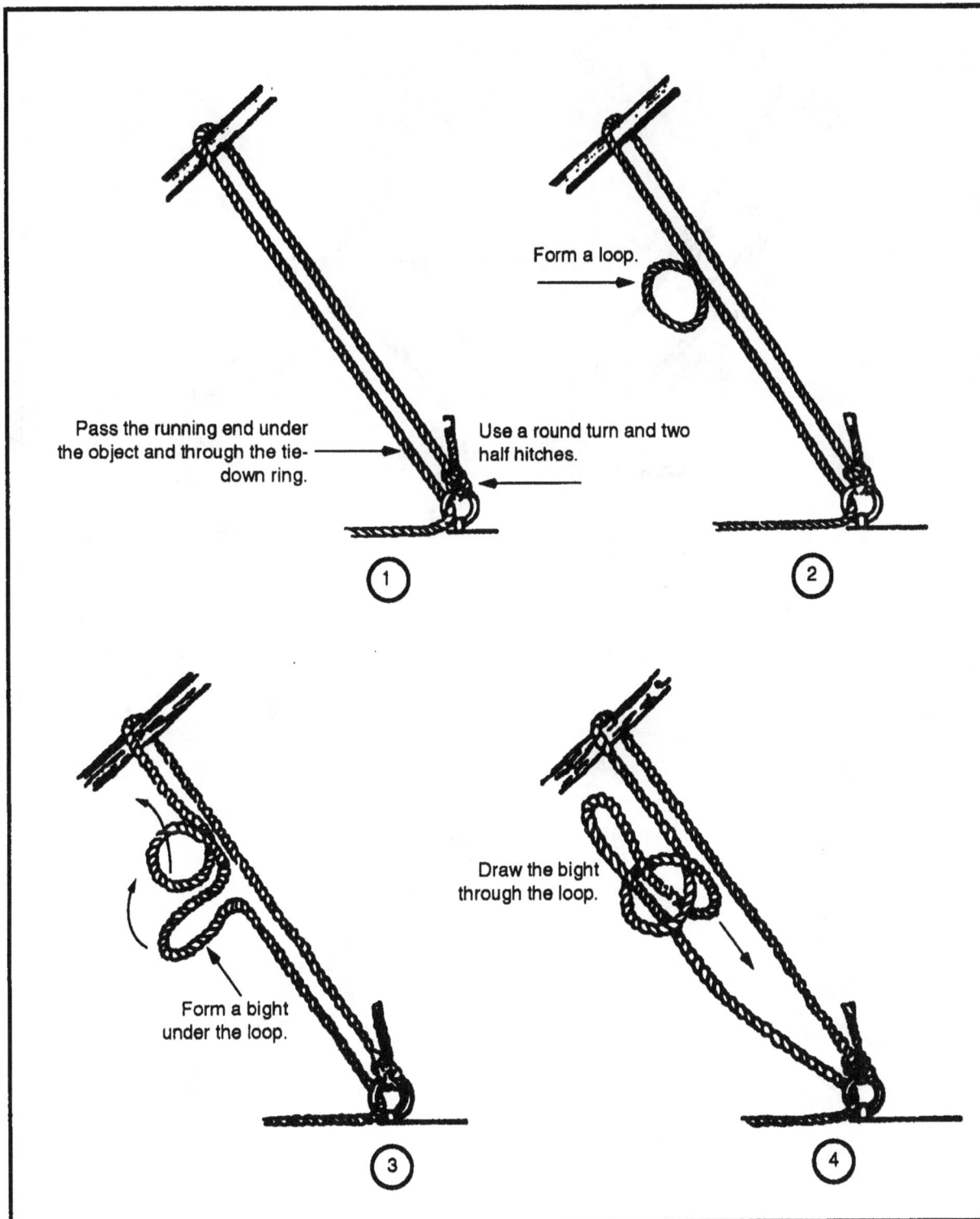

Figure 2-21. Baker bowline

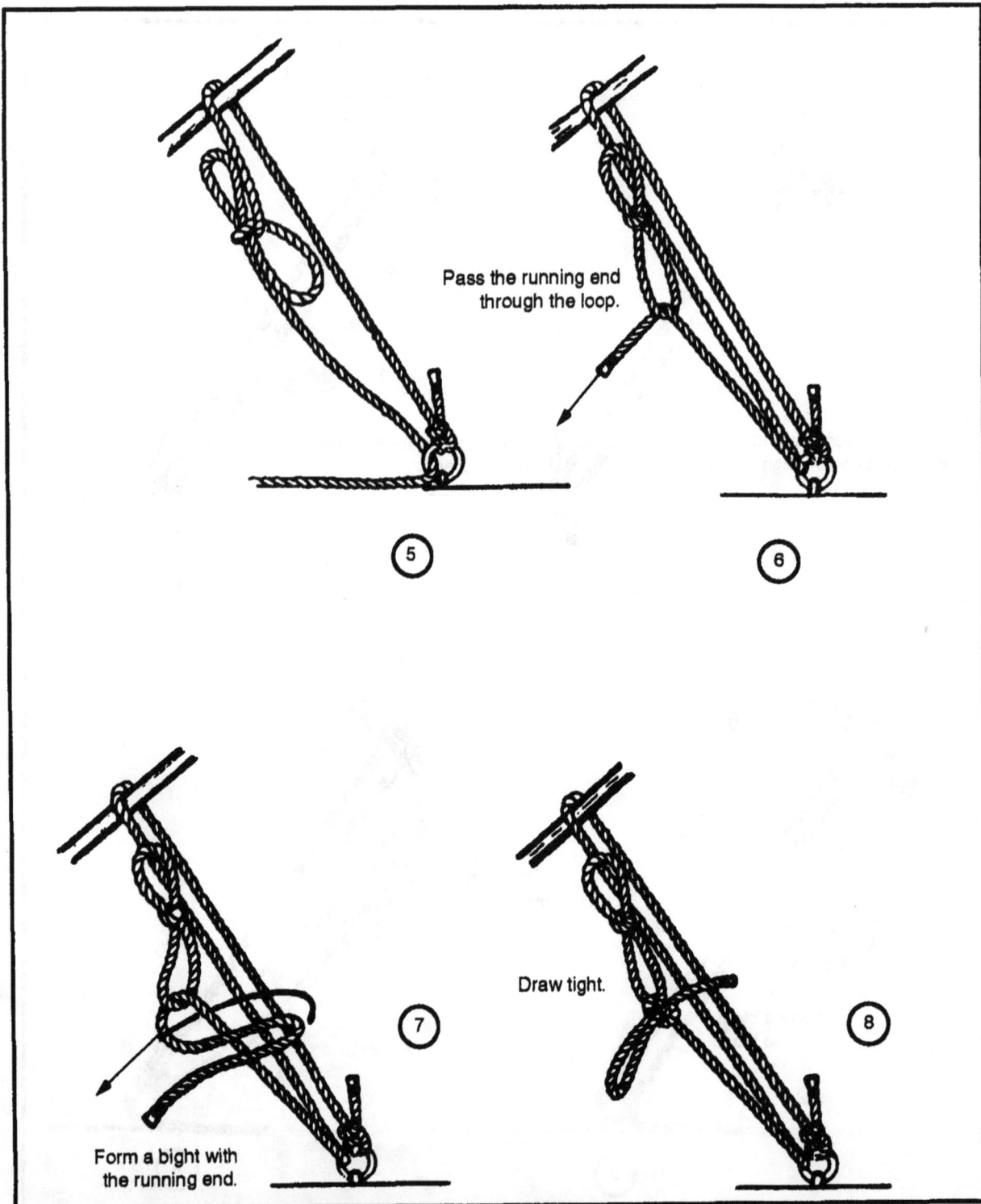

Pass the running end
through the loop.

⑤

⑥

Form a bight with
the running end.

⑦

Draw tight.

⑧

Figure 2-21. Baker bowline (continued)

wire or strands of cordage. Check all knots in wire rope periodically for wear or signs of breakage. If there is any reason to believe that the knot has been subjected to excessive wear, cut off a short length of the end of the rope, including the knot, and tie a new knot. Use the fisherman's bend, clove hitch, and carrick bend to fasten wire rope.

HITCHES

A hitch is any of various knots used to form a temporary noose in a rope or to secure a rope around a timber, pipe, or post so that it will hold temporarily but can be readily undone. The types of hitches are as follows:

- Half hitch.
- Two half hitches.
- Round turn and two half hitches.
- Timber hitch.
- Timber hitch and half hitch.
- Clove hitch.
- Rolling hitch.
- Telegraph hitch.
- Mooring hitch.
- Scaffold hitch.
- Blackwall hitch.
- Harness hitch.
- Girth hitch.
- Sheepshank.
- Fisherman's bend.

HALF HITCH

Use the half hitch to tie a rope to a timber or to a larger rope (see *Figure 2-22, A*). It will hold against a steady pull on the standing part of the rope; however, it is not a secure hitch. You can use the half hitch to secure the free end of a rope and as an aid to and the foundation of many knots. For example, it is the start of a timber hitch and a part of the fisherman's knot. It also makes the rolling hitch more secure.

TWO HALF HITCHES

Two half hitches are especially useful for securing the running end of a rope to the standing part (see *Figure 2-22, B*). If the two hitches are slid together along the standing part to form a single knot, the knot becomes a clove hitch.

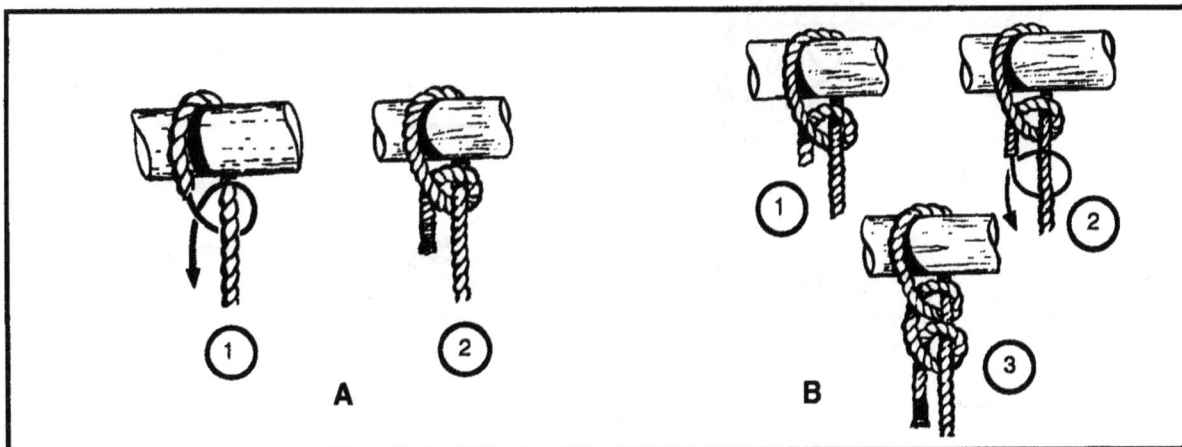

Figure 2-22. Half hitches

ROUND TURN AND TWO HALF HITCHES

Another hitch used to fasten a rope to a pole, timber, or spar is the round turn and two half hitches (see *Figure 2-23*).For greater security, seize the running end of the rope to the standing part. This hitch does not jam.

TIMBER HITCH

Use the timber hitch to move heavy timber or poles (see *Figure 2-24*)It is excellent for securing a piece of lumber or similar objects. The pressure of the coils, one over the other, holds the timber securely; the more tension applied, the tighter the hitch becomes about the timber. It will not slip but will readily loosen when the strain is relieved.

TIMBER HITCH AND HALF HITCH

A timber hitch and half hitch are combined to hold heavy timber or poles when they are being lifted or dragged (see *Figure 2-25*)A timber hitch used alone may become untied when the rope is slack or when a sudden strain is put on it.

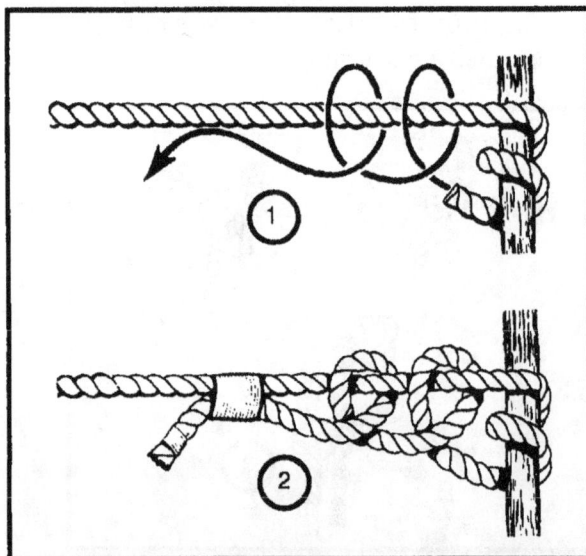

Figure 2-24. Timber hitch

CLOVE HITCH

The clove hitch is one of the most widely used knots (see*Figure 2-26, page 2-19*)You can use it to fasten a rope to a timber, pipe, or post. You can also use it to make other knots. This knot puts very little strain on the fibers when the rope is put around an object in one continuous direction. You can tie a clove hitch at any point in a rope. If there is not constant tension on the rope, another loop (round of the rope around the object and under the center of the clove hitch) will permit a tightening and slackening motion of the rope.

ROLLING HITCH

Use the rolling hitch to secure a rope to another rope or to fasten it to a pole or pipe

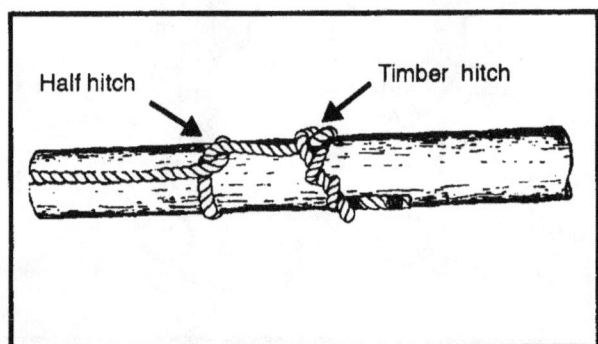

Figure 2-25. Timber hitch and half hitch

Figure 2-23. Round turn and two half hitches

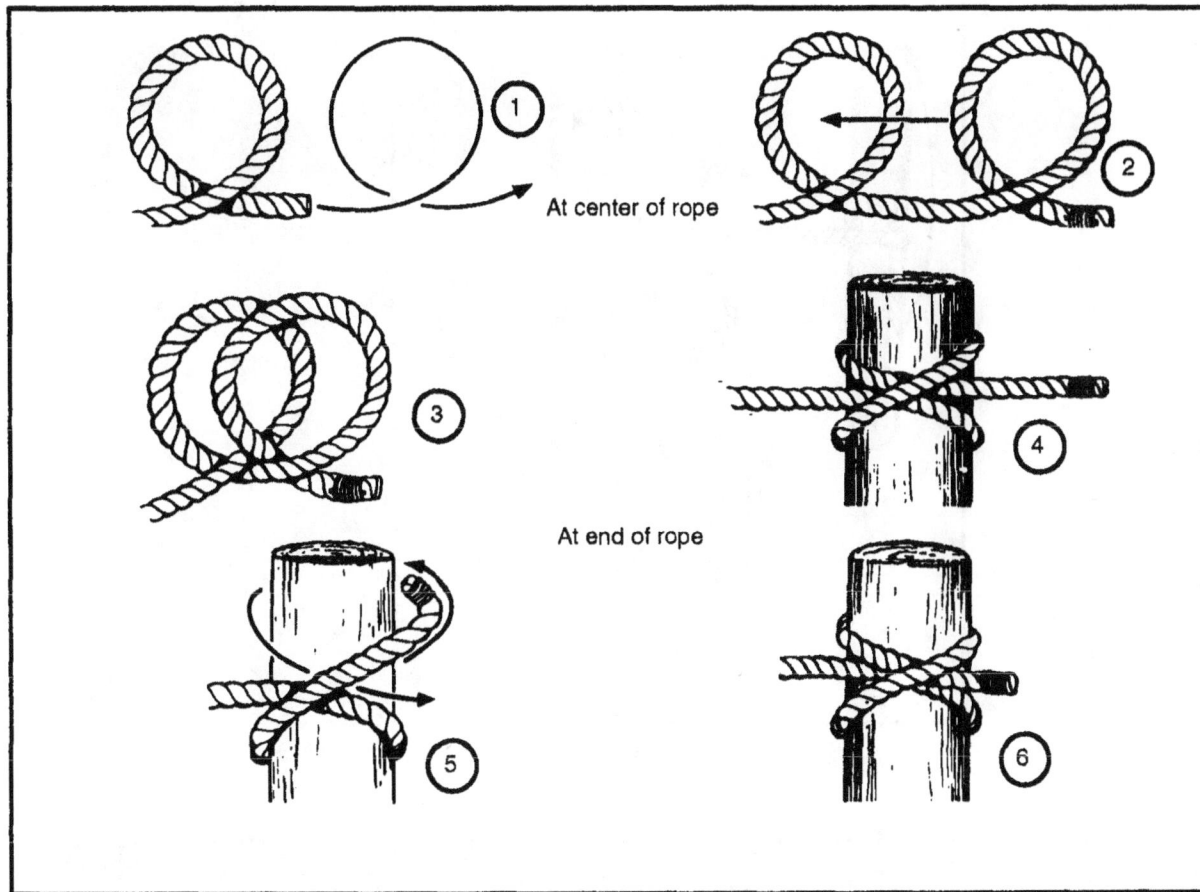

Figure 2-26. Clove hitch

so that the rope will not slip (see *Figure 2-27, page 2-20*). This knot grips tightly but is easily moved along a rope or pole when the strain is relieved.

TELEGRAPH HITCH

The telegraph hitch is a very useful and secure hitch that you can use to hoist or haul posts and poles (see *Figure 2-28, page 2-20*). It is easy to tie and untie and will not slip.

MOORING HITCH

Use the mooring hitch, also called rolling or magnus hitch, to fasten a rope around a mooring post or to attach a rope at a right angle to a post (see *Figure 2-29, page 2-21*).

This hitch grips tightly and is easily removed.

SCAFFOLD HITCH

Use the scaffold hitch to support the end of a scaffold plank with a single rope (see *Figure 2-30, page 2-21*). It prevents the plank from tilting.

BLACKWALL HITCH

Use the blackwall hitch to fasten a rope to a hook (see *Figure 2-31, page 2-22*). Generally, use it to attach a rope, temporarily, to a hook or similar object in derrick work. The hitch holds only when subjected to a constant strain or when used in the middle of a

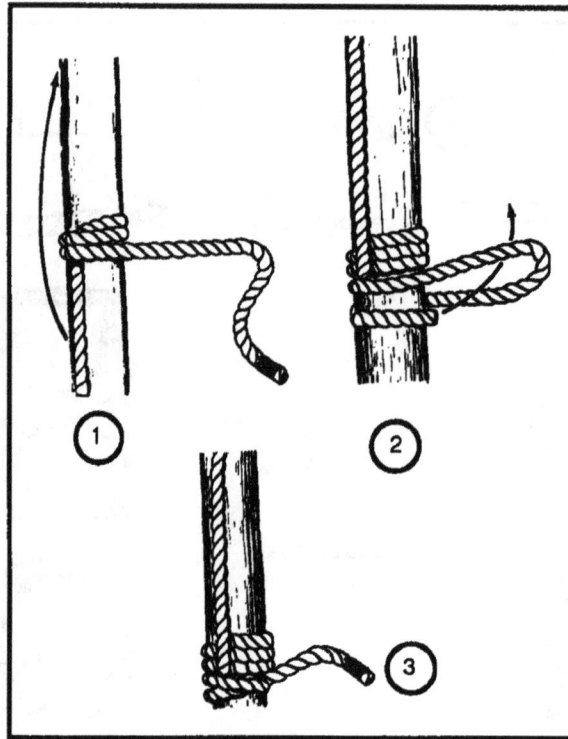

Figure 2-27. Rolling hitch

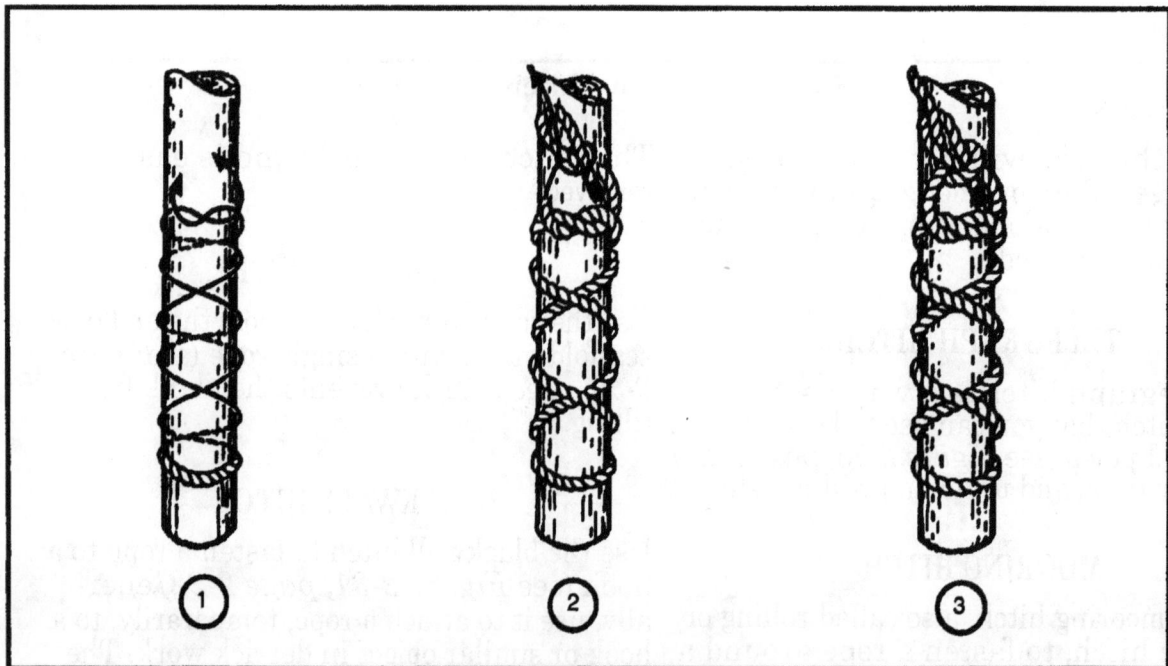

Figure 2-28. Telegraph hitch

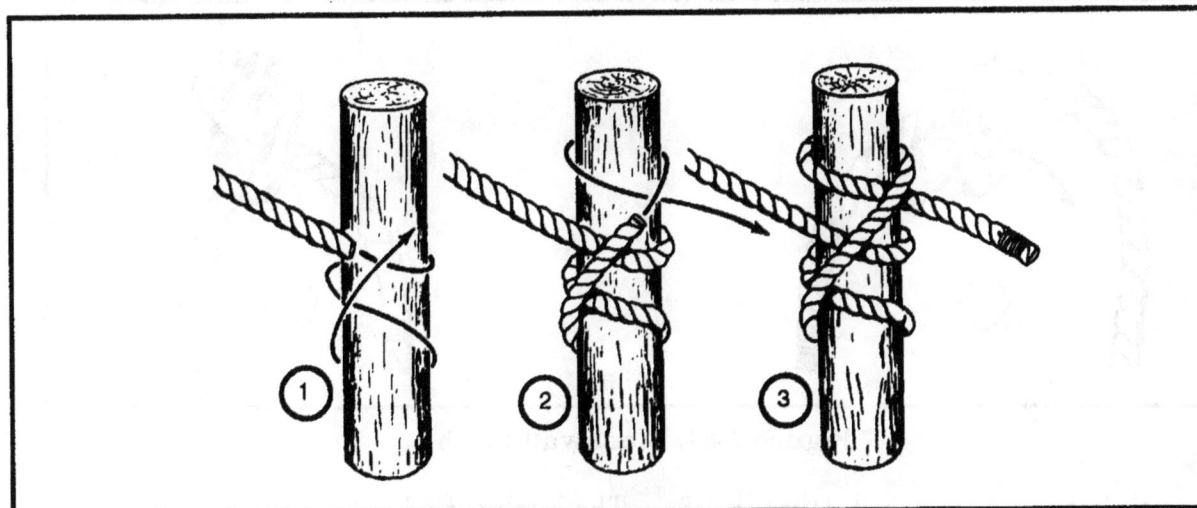

Figure 2-29. Mooring hitch

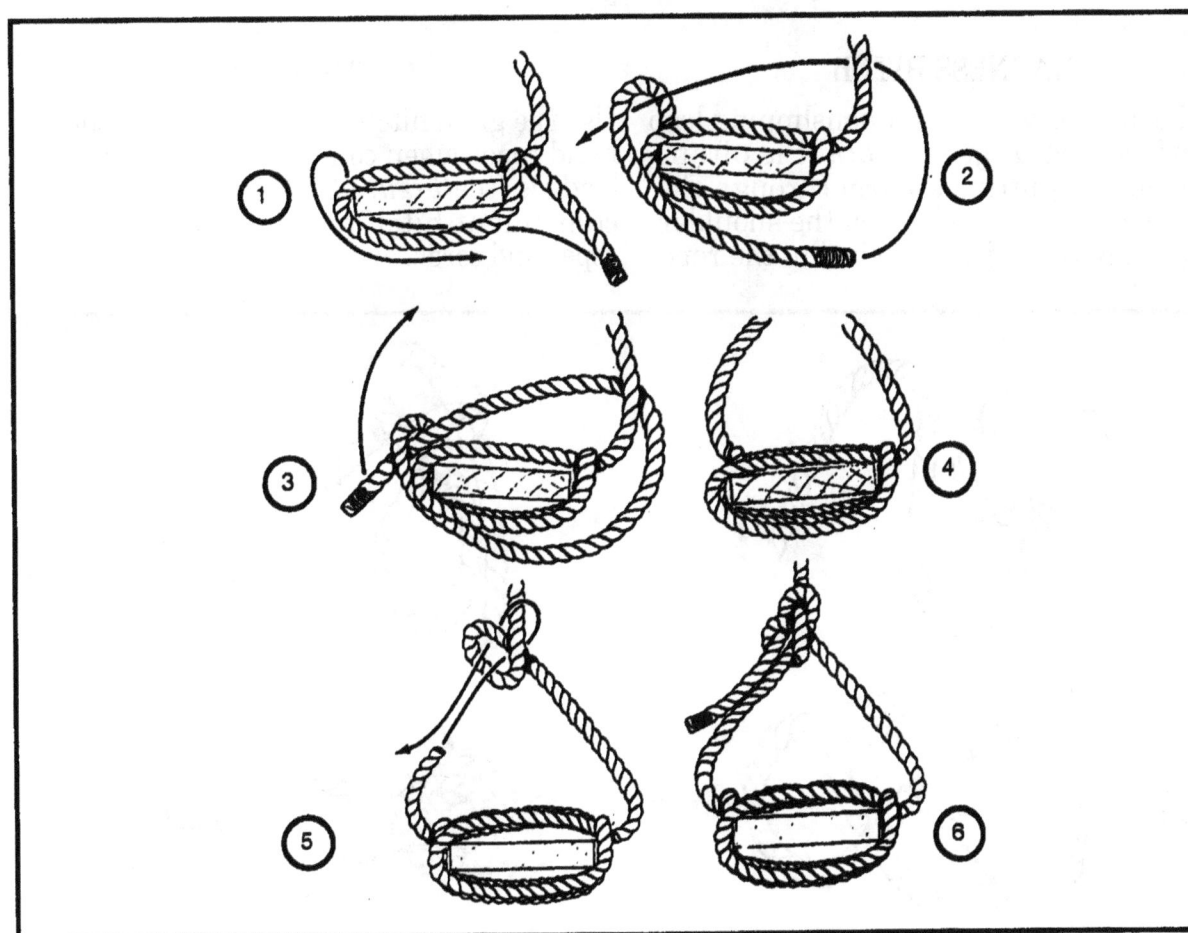

Figure 2-30. Scaffold hitch

Figure 2-31. Blackwall hitch

rope with both ends secured. Human life and breakable equipment should never be entrusted to the blackwall hitch.

The hitch is tied only in the middle of a rope. It will slip if only one end of the rope is pulled.

HARNESS HITCH

The harness hitch forms a nonslipping loop in a rope (see *Figure 2-32*). It is often employed by putting an arm through the loop, then placing the loop on the shoulder and pulling the object attached to the rope.

GIRTH HITCH

Use the girth hitch to tie suspender ropes to hand ropes when constructing expedient foot bridges (see *Figure 2-33*). It is a simple and convenient hitch for many other uses of ropes and cords.

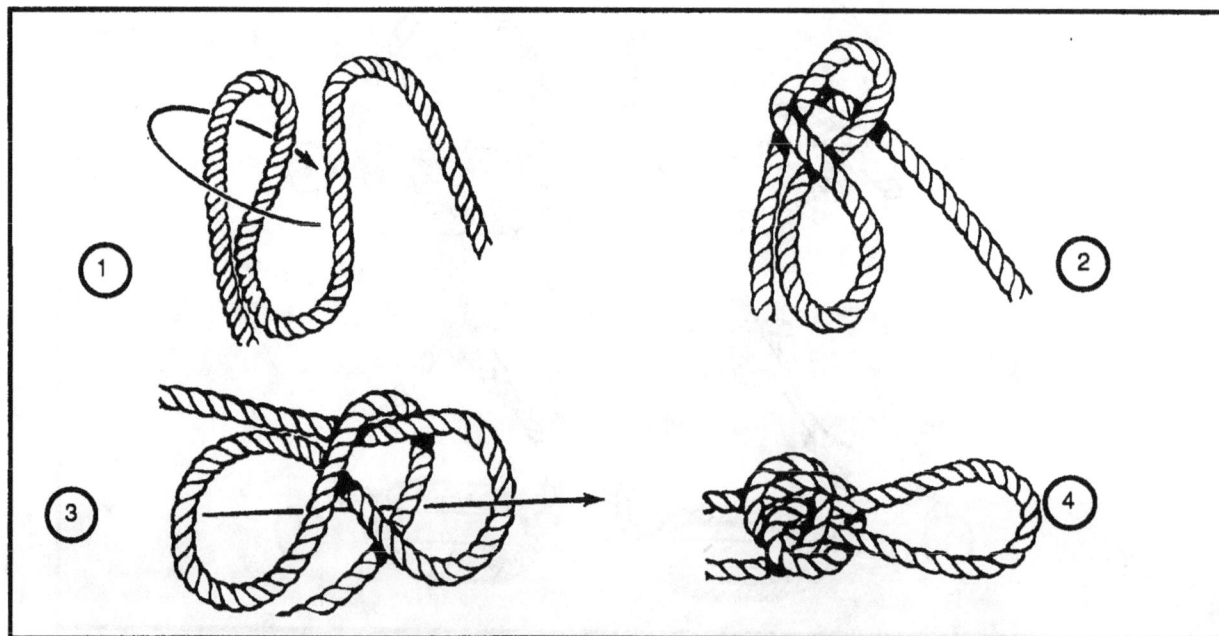

Figure 2-32. Harness hitch

Figure 2-33. Girth hitch

SHEEPSHANK

A sheepshank is a method of shortening a rope, but you can use it to take the load off a weak spot in the rope (see *Figure 2-34*) It is only a temporary knot unless the eyes are fastened to the standing part on each end.

FISHERMAN'S BEND

The fisherman's bend is an excellent knot for attaching a rope to a light anchor, a ring, or a rectangular piece of stone (see *Figure 2-35, page 2-24*) You can use it to fasten a rope or cable to a ring or post. Also use it where there will be a slackening and tightening motion in the rope.

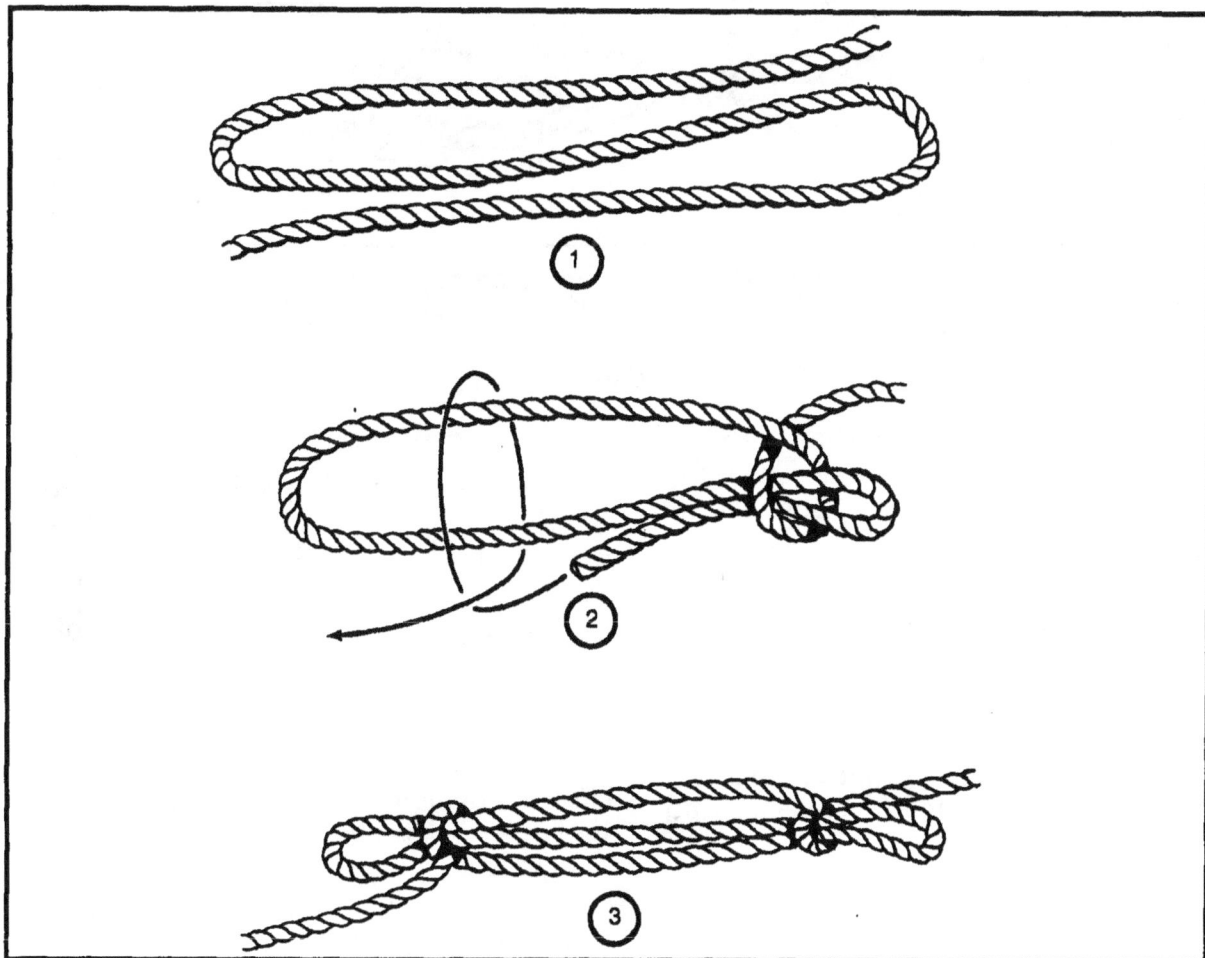

Figure 2-34. Sheepshank

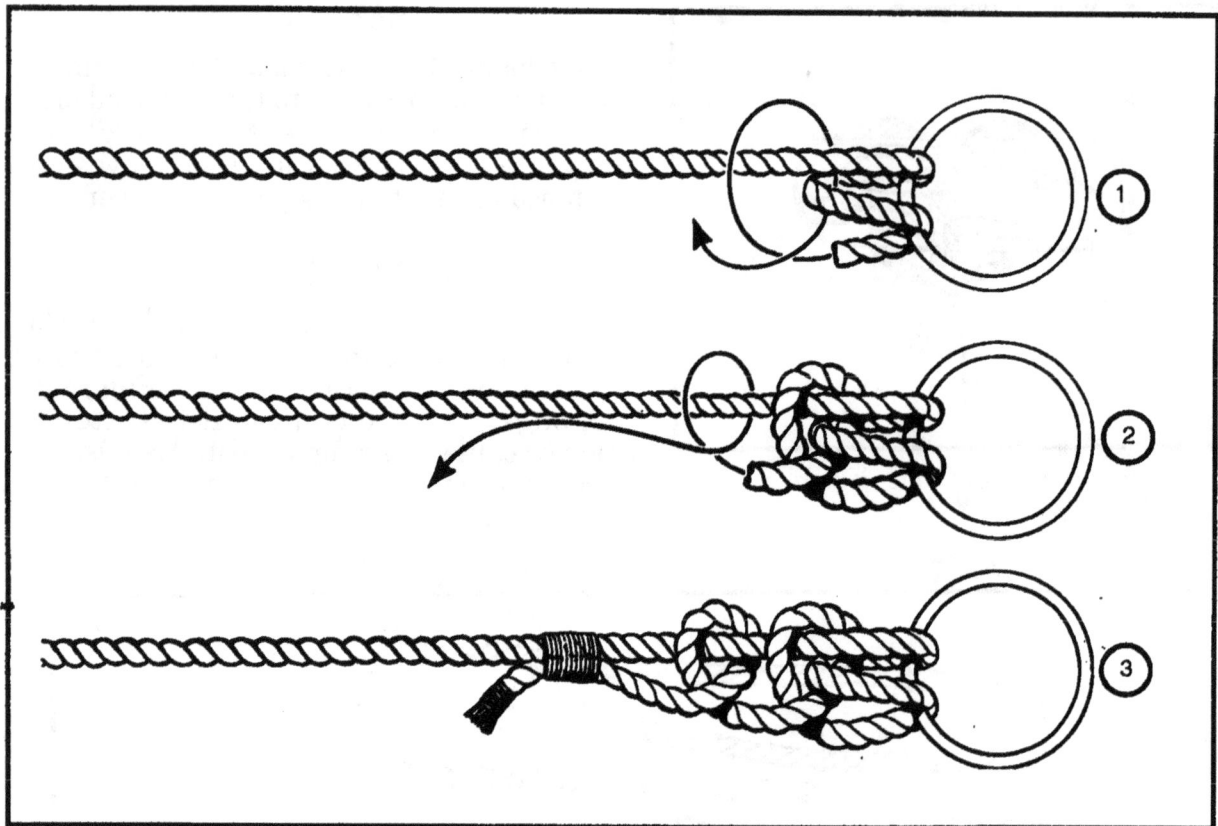

Figure 2-35. Fisherman's bend

LASHINGS

A lashing is as rope, wire, or chain used for binding, wrapping, or fastening. The types of lashings include square, shears, and block.

SQUARE LASHING

Use the square lashing to lash two spars together at right angles to each other (see *Figure 2-36).*To tie a square lashing, begin with a clove hitch on one spar and make a minimum of four complete turns around both members. Continue with two frapping turns between the vertical and the horizontal spar to tighten the lashing. Tie off the running end to the opposite spar from which you started with another clove hitch to finish the square lashing.

SHEARS LASHING

Use the shears lashing to lash two spars together at one end to form an expedient device called a shears (see*Figure 2-37).*Do this by laying two spars side by side, spaced about one-third of the diameter of a spar apart, with the butt ends together. Start the shears lashing a short distance in from the top of one of the spars by tying the end of the rope to it with a clove hitch. Then make eight tight turns around both spars above the clove hitch. Tighten the lashing with a minimum of two frapping turns around the eight turns. Finish the shears lashing by tying the end of the rope to the opposite spar from which you started with another clove hitch.

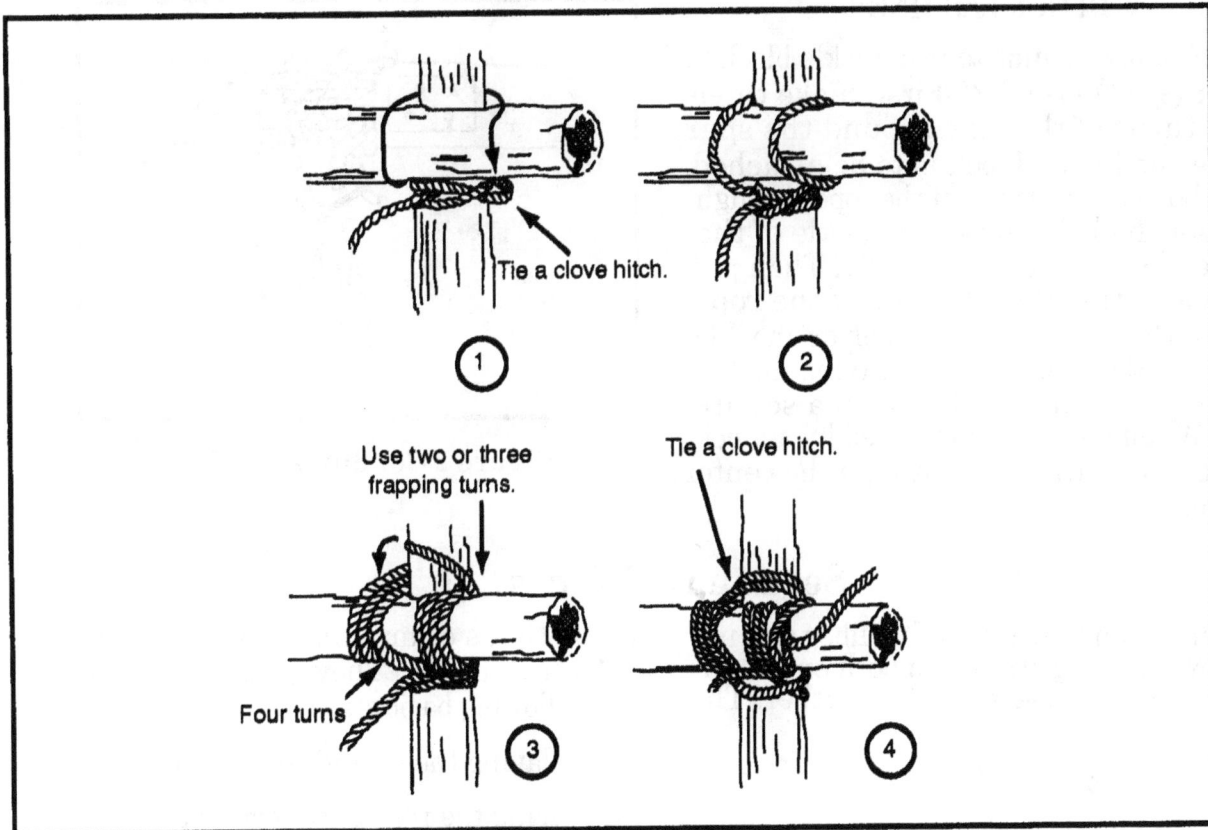

Tie a clove hitch.

① ②

Use two or three
frapping turns.

Tie a clove hitch.

Four turns

③ ④

Figure 2-36. Square lashing

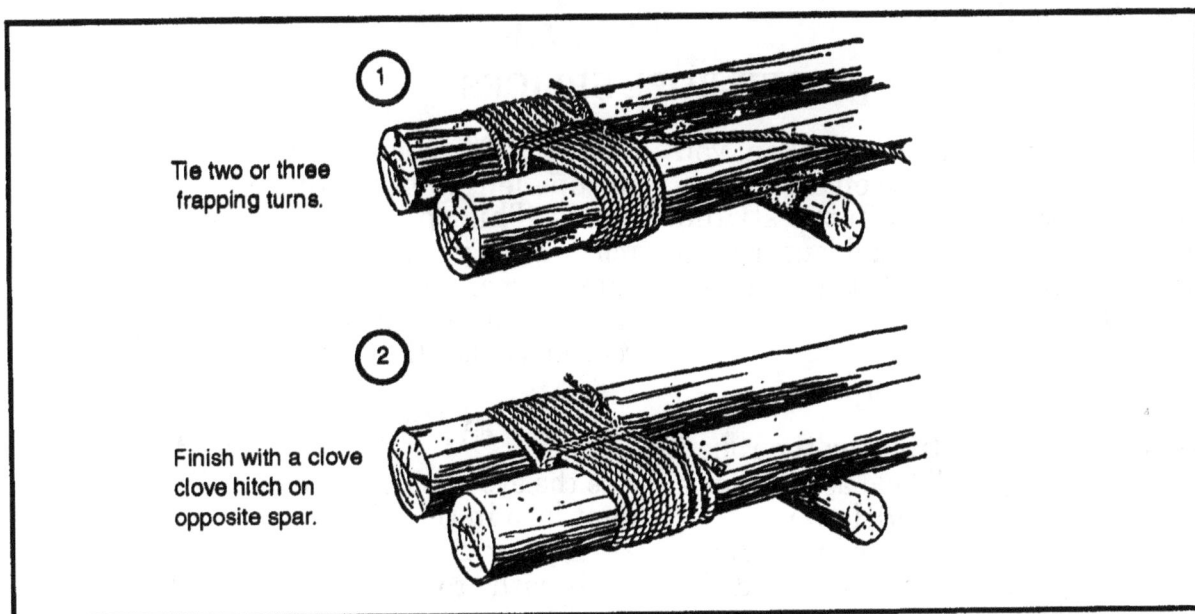

①

Tie two or three
frapping turns.

②

Finish with a clove
clove hitch on
opposite spar.

Figure 2-37. Shears lashing

BLOCK LASHING

Use the block lashing to tie a tackle block to a spar (see *Figure 2-38*)First, make three right turns of the rope around the spar where the tackle block is to be attached. Pass the next two turns of the rope through the mouth of the hook or shackle of the tackle block and drawn tightly. Then put three additional taut turns of the rope around the spar above the hook or shackle. Complete the block lashing by tying the two ends of the rope together with a square knot. When a sling is supported by a block lashing, pass the sling through the center four turns.

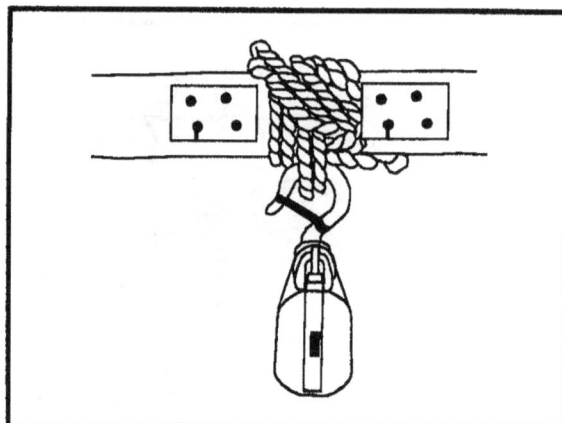

Figure 2-38. Block lashing

Section II. Splices

Splicing is a method of joining fiber or wire rope by unlaying strands of both ends and interweaving these strands together. The general types of splices are—

- A short splice.
- An eye or side splice.
- A long splice.
- A crown or back splice.

The methods of making all four types of splices are similar. They generally consist of the following basic steps—

- Unlaying the strands of the rope.
- Placing the rope ends together.
- Interweaving the strands and tucking them into the rope.

FIBER-ROPE SPLICES

When one strand of a rope is broken, you cannot repair it by tying the ends together because this would shorten the strand. Repair it by inserting a strand longer than the break and tying the ends together (see *Figure 2-39*).

SHORT SPLICE

The short splice is as strong as the rope in which it is made and will hold as much as a long splice (see *Figure 2-40*)However, the short splice causes an increase in the diameter of the rope for a short distance and can be used only where this increase in diameter will not affect operations. It is called the

short splice because a minimum reduction in rope length takes place in making the splice. This splice is frequently used to repair damaged ropes when two ropes of the same size are to be joined together permanently. Cut out the damaged parts of the rope and splice the sound sections.

EYE OR SIDE SPLICE

Use the eye or side splice to make a permanent loop in the end of a rope (see *Figure 2-41, page 2-28*)You can use the loops, made with or without a thimble, to fasten the rope to a ring or hook. Use a thimble to reduce wear. Use this splice also to splice

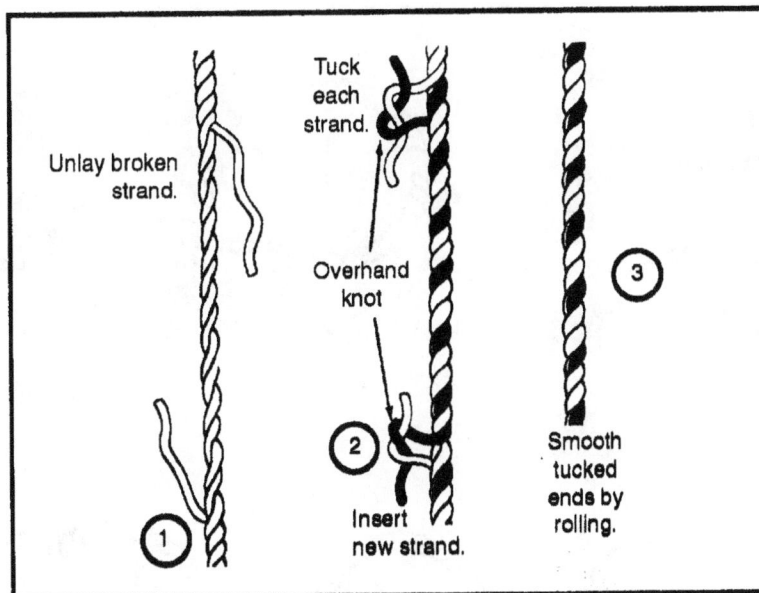

Figure 2-39. Renewing rope strands

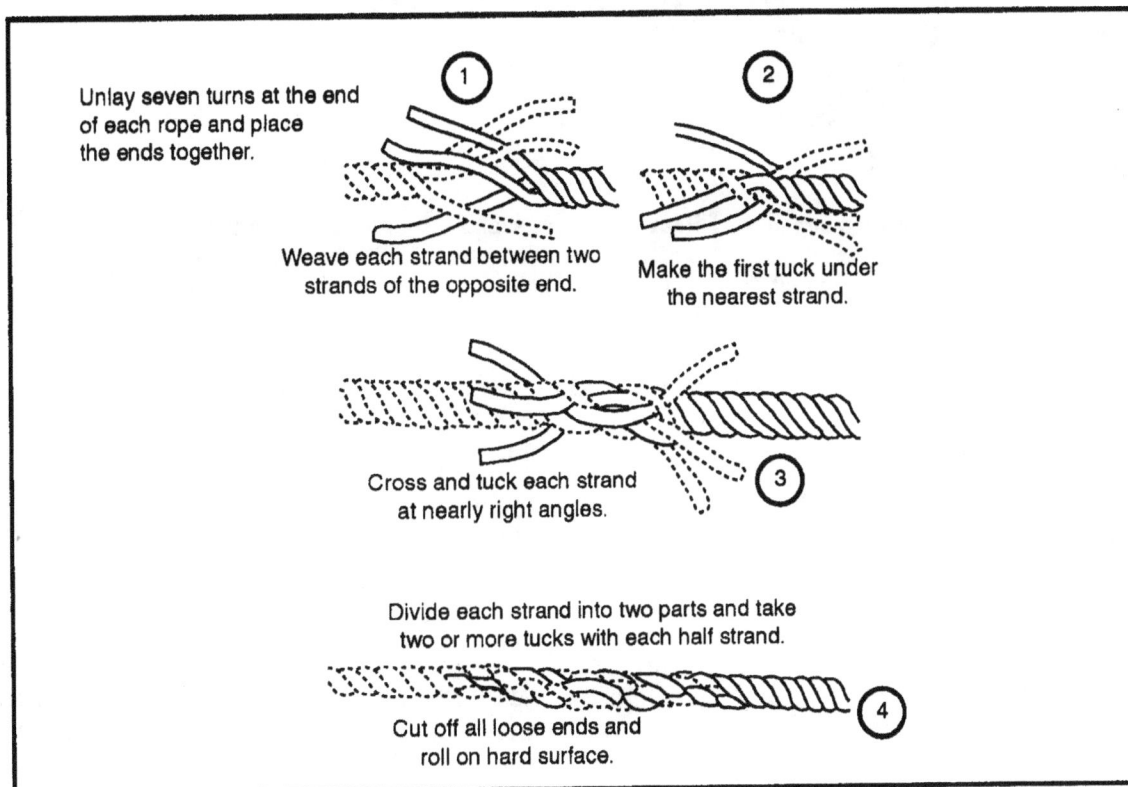

Figure 2-40. Short splice for fiber rope

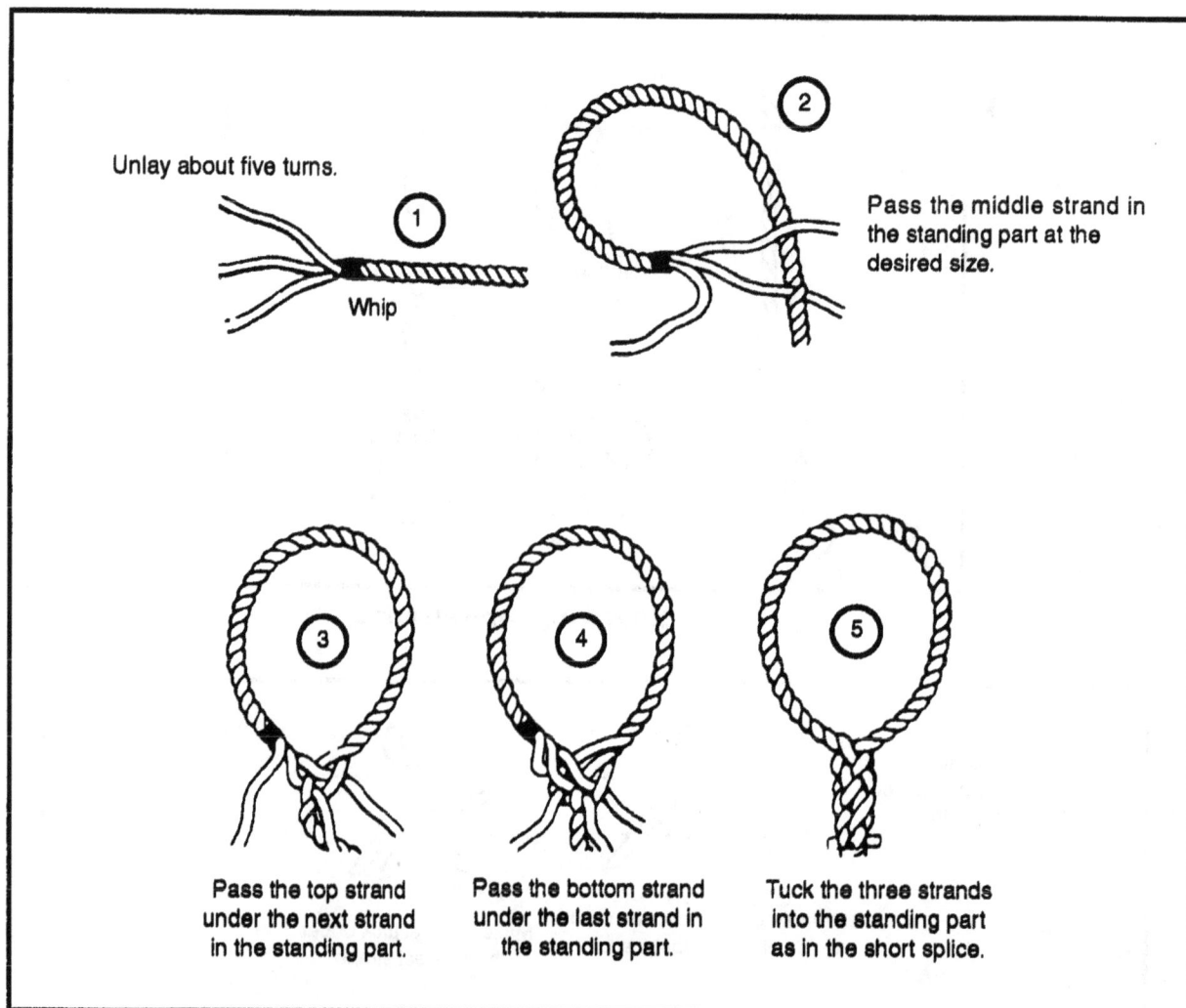

Figure 2-41. Eye or side splice for fiber rope

one rope into the side of another. As a permanent loop or eye, no knot can compare with this splice for neatness and efficiency.

The ropes to be joined should be the same lay and as nearly the same diameter as possible.

LONG SPLICE

Use the long splice when the larger diameter of the short splice has an adverse effect on the use of the rope; use it also to splice long ropes that operate under heavy stress (see *Figure 2-42)*.This splice is as strong as the rope itself. A skillfully made long splice will run through sheaves without any difficulty.

CROWN OR BACK SPLICE

When you are splicing the end of a rope to prevent unlaying, and a slight enlargement of the end is not objectionable, use a crown splice to do this (see*Figure 2-43, page 2-30)*.Do not put any length of rope into service without properly preparing the ends.

Unlay fifteen turns from each end.

Unlay one strand and
lay in its place a strand
of the other rope.

Bring the ropes together
as in the short splice.

Leave five turns.

Be sure the ends of the strand in each
pair pass each other.

Tuck and finish each pair as
in the short splice.

Cut off all loose ends.

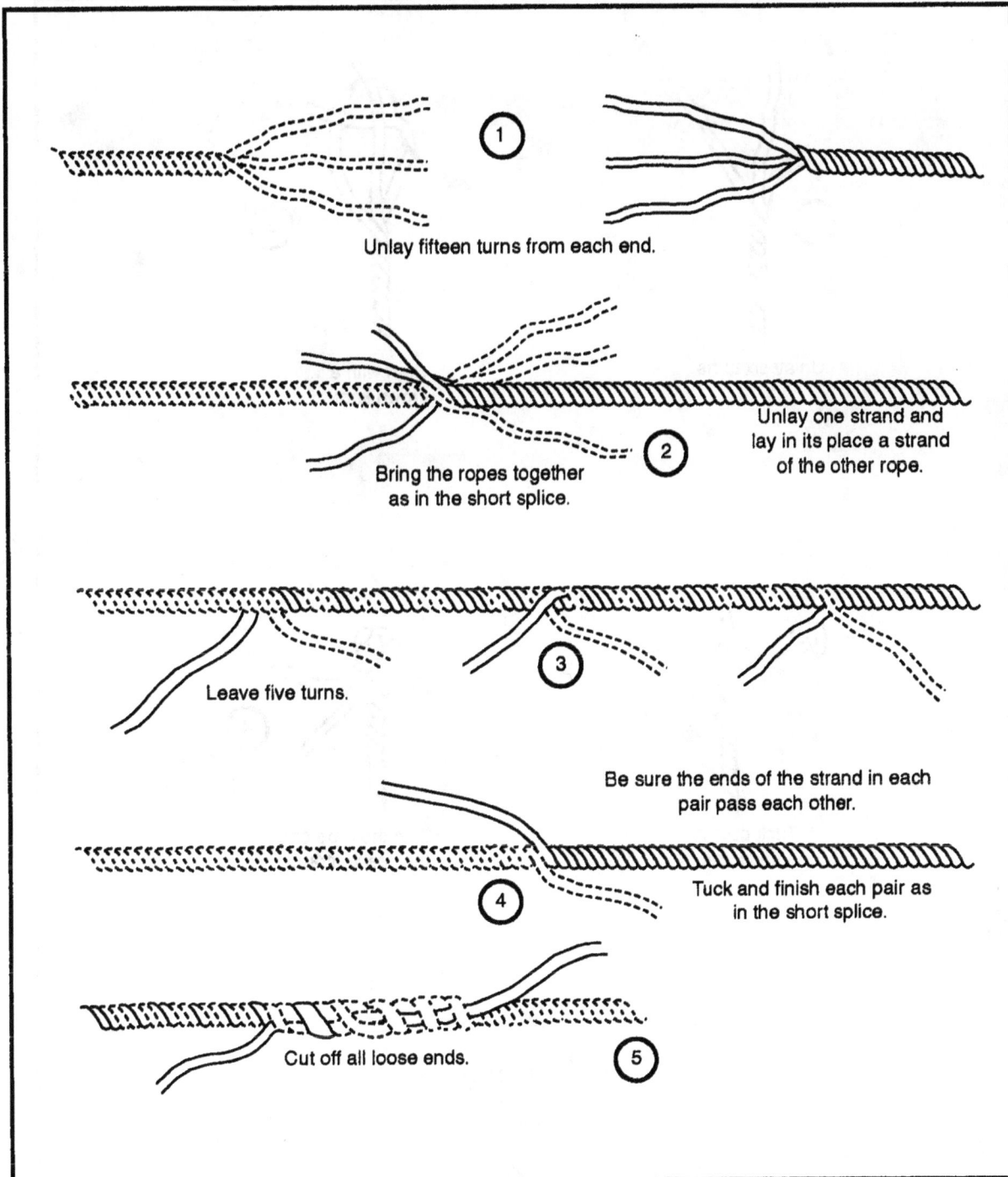

Figure 2-42. Long splice for fiber rope

1. Unlay six turns.

2. Start with a crown knot.

3. Tuck over one and under the next.

4. Turn the rope and tuck each strand.

5. Trim the ends.

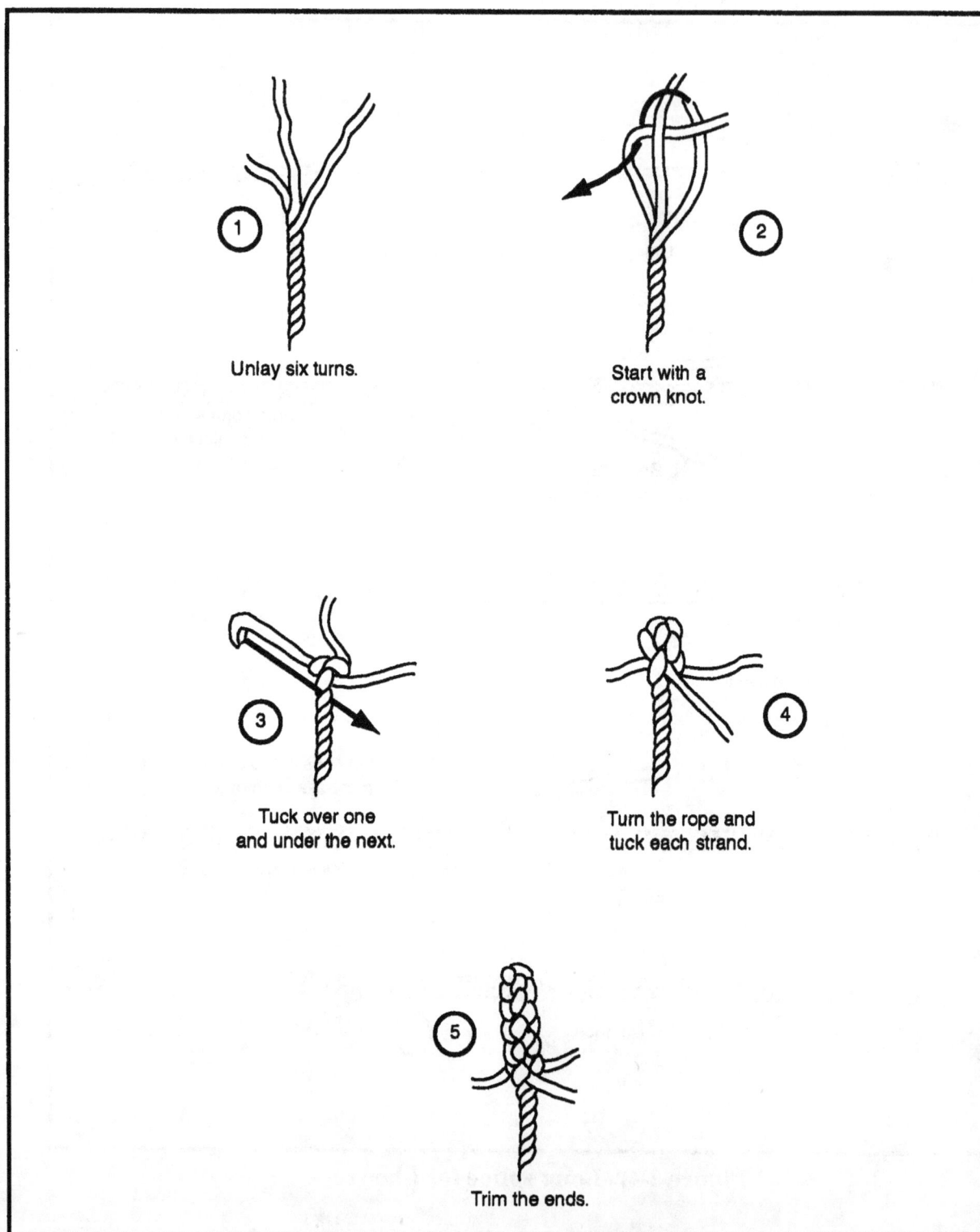

Figure 2-43. Crown or back splice for fiber rope

WIRE-ROPE SPLICES

In splicing wire rope, it is extremely important to use great care in laying the various rope strands firmly into position. Slack strands will not receive their full share of the load, which causes excessive stress to be put on the other strands. The unequal stress distribution will decrease the possible ultimate strength of the splice. When using splices in places where their failure may result in material damage or may endanger human lives, test the splices under stresses equal to at least twice their maximum working load before placing the ropes into service. Table 2-2 shows the amount or length of rope to be unlaid on each of the two ends of the ropes and the amount of tuck for ropes of different diameters. As a rule of thumb, use the following:

- Long splice, 40 times the diameter.
- Short splice, 20 times the diameter.

You need only a few tools to splice wire rope. In addition to the tools shown in *Figure 2-44, page 2-32*, a hammer and cold chisel are often used to cut the ends of strands. Use two slings of marline and two sticks to

untwist the wire. A pocket knife may be needed to cut the hemp core.

SHORT SPLICE

A short splice develops only from 70 to 90 percent of the strength of the rope. Since a short splice is bulky, it is used only for block straps, slings, or where an enlargement of the diameter is of no importance. It is not suitable for splicing driving ropes or ropes used in running tackles and should never be put into a crane or hoist rope. The wire rope splice differs from the fiber rope short splice only in the method by which the end strands are tucked (see *Figure 2-45, page 2-32*).

EYE OR SIDE SPLICE

An eye splice can be made with or without a thimble. Use a thimble for every rope eye unless special circumstances prohibit it (see *Figure 2-46, page 2-33*). The thimble protects the rope from sharp bends and abrasive action. The efficiency of a well-made eye splice with a heavy-duty thimble varies

Table 2-2. Amount of wire rope to allow for splices and tucks

Diameter (Inches)	Length of Rope to Allow (feet)			Tuck Length (inches)		
	Short Splice	Eye Splice	Long Splice	Short Splice	Eye Splice	Long Splice
1/4-3/8	15	1	30	15	1	30
1/2 – 5/8	20	2	40	20	2	40
3/4 – 7/8	24	2 1/2	50	24	2 1/2	50
1 –1 1/8	28	3	60	28	3	60
1 1/4 – 1 3/8	32	3 1/2	70	32	3 1/2	70
1 1/2	36	4	80	36	4	80

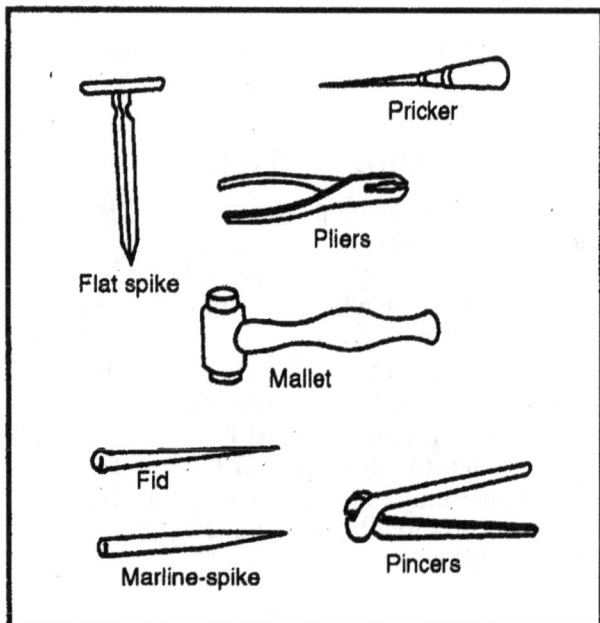

Figure 2-44. Tools for wire splicing

from 70 to 90 percent. Occasionally, it becomes necessary to construct a field expedient, called a hasty eye (see *Figure 2-47*). The hasty eye can be easily and quickly made but is limited to about 70 percent of the strength of the rope; consequently, it should not be used to hoist loads.

LONG SPLICE

Use the long splice to join two ropes or to make an endless sling without increasing the thickness of the wire rope at the splice (see *Figure 2-48, page 2-34*). It is the best and most important kind of splice because it is strong and trim.

Round-Strand, Regular-Lay Rope

The directions given in *Figure 2-48* are for making a 30-foot splice in a three-fourths

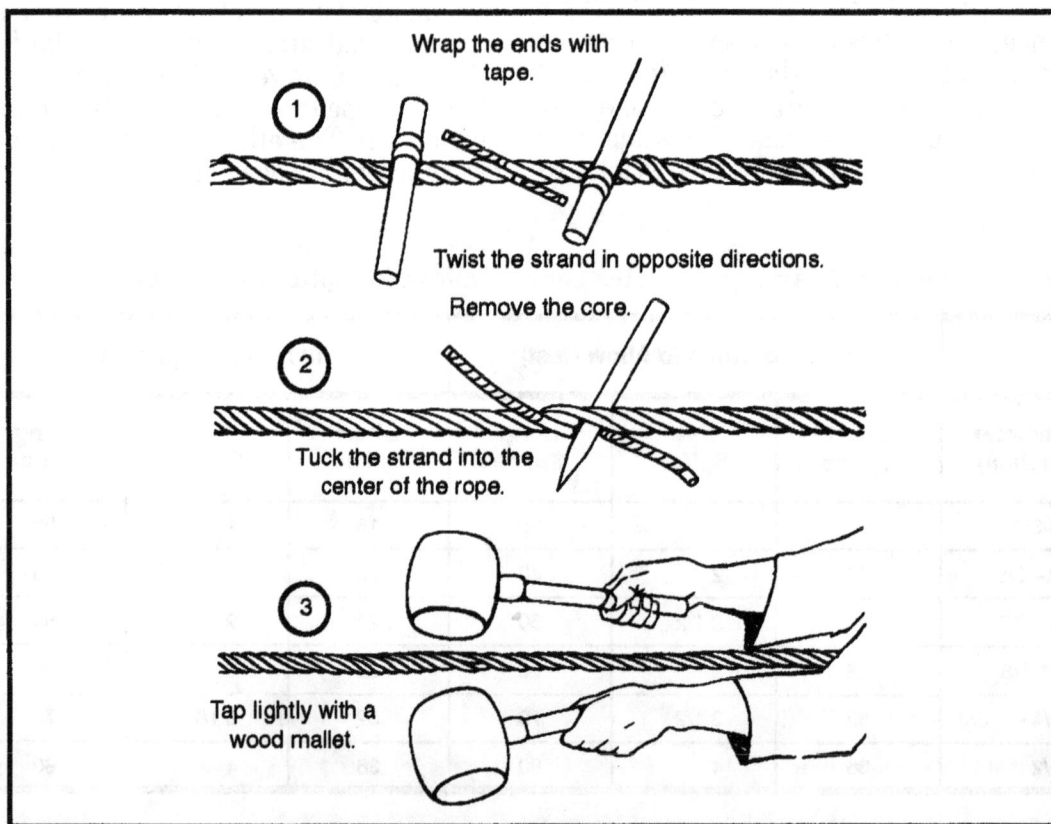

Figure 2-45. Tucking wire-rope strands

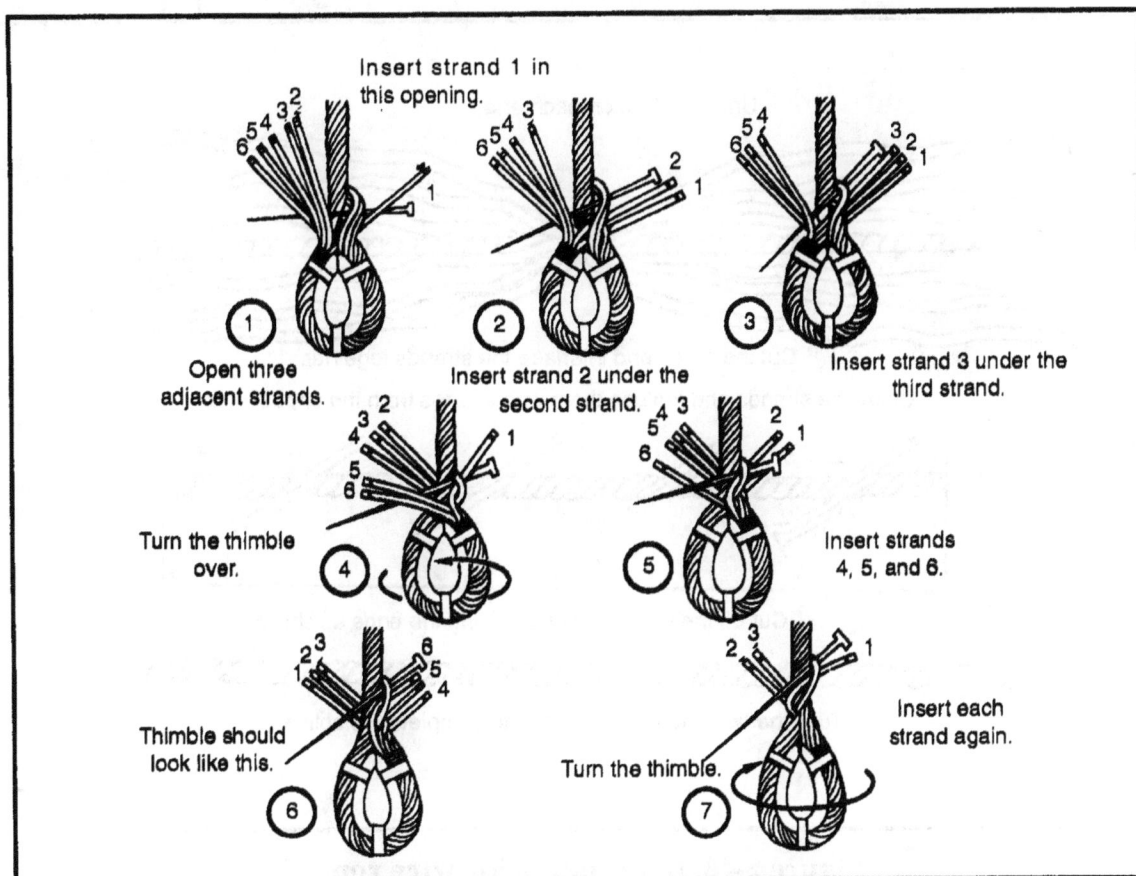

Figure 2-46. Eye splice with thimble for wire rope

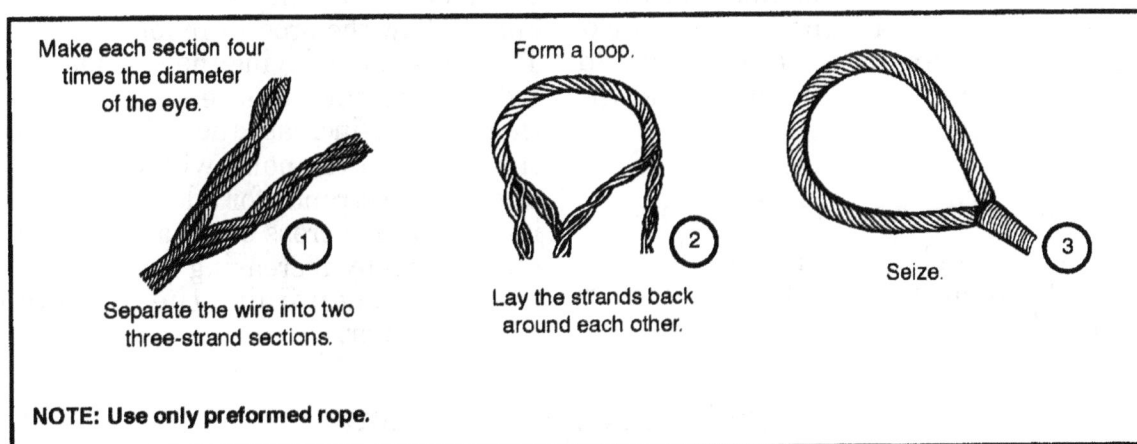

Figure 2-47. Hasty eye splice for wire rope

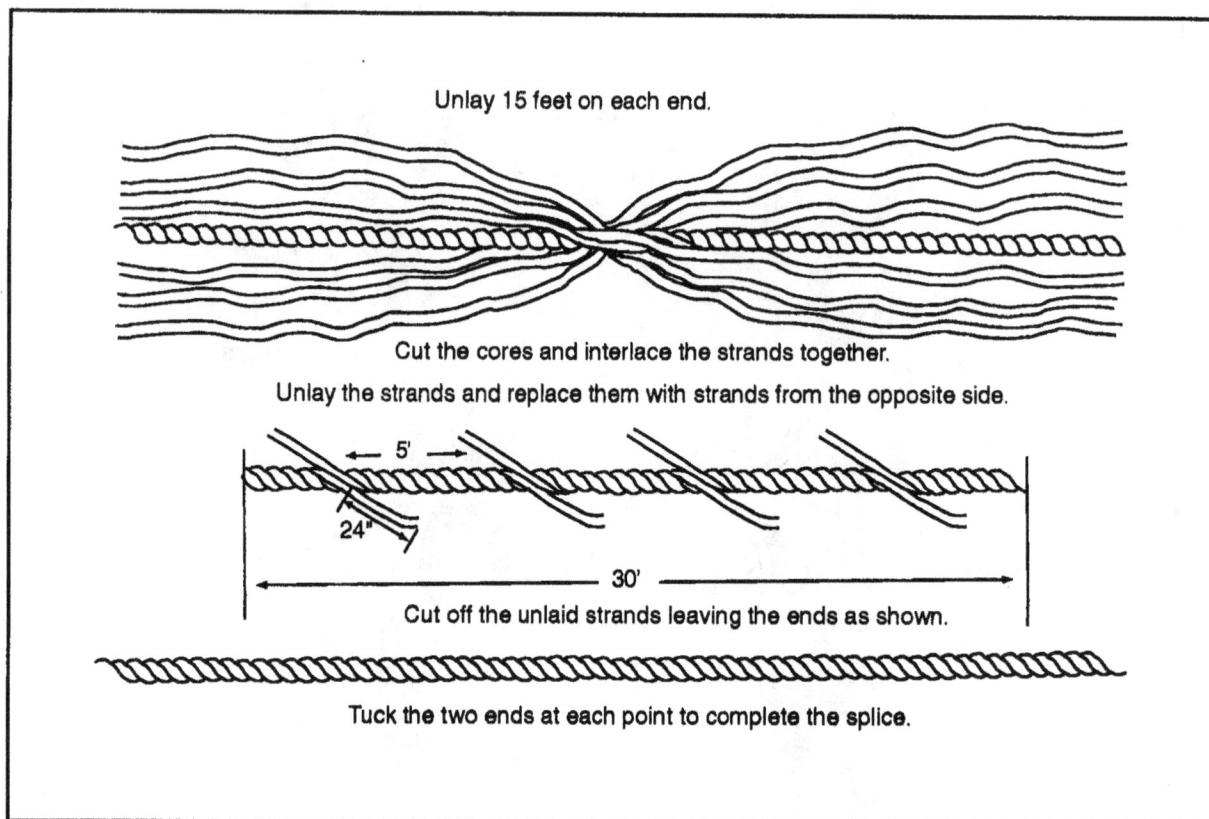

Unlay 15 feet on each end.

Cut the cores and interlace the strands together.

Unlay the strands and replace them with strands from the opposite side.

5'

24"

30'

Cut off the unlaid strands leaving the ends as shown.

Tuck the two ends at each point to complete the splice.

Figure 2-48. Long splice for wire rope

inch regular-lay, round-strand, hemp-center wire rope. Other strand combinations differ only when there is an uneven number of strands. In splicing ropes having an odd number of strands, make the odd tuck at the center of the splice.

Round-Strand, Lang-Lay Rope

In splicing a round-strand, Lang-lay rope, it is advisable to make a slightly longer splice than for the same size rope of regular lay

because of the tendency of the rope to untwist. Up to the point of tucking the ends, follow the procedure for regular lay. Then, instead of laying the strands side by side where they pass each other, cross them over to increase the holding power of the splice. At the point where they cross, untwist the strands for a length of about 3 inches so they cross over each other without materially increasing the diameter of the rope. Then finish the tucks in the usual manner.

Section III. Attachments

Most of the attachments used with wire rope are designed to provide an eye on the end of the rope by which maximum strength can be obtained when the rope is connected with another rope, hook, or ring. *Figure 2-49* shows

a number of attachments used with the eye splice. Any two of the ends can be joined together, either directly or with the aid of a shackle or end fitting. These attachments for wire rope take the place of knots.

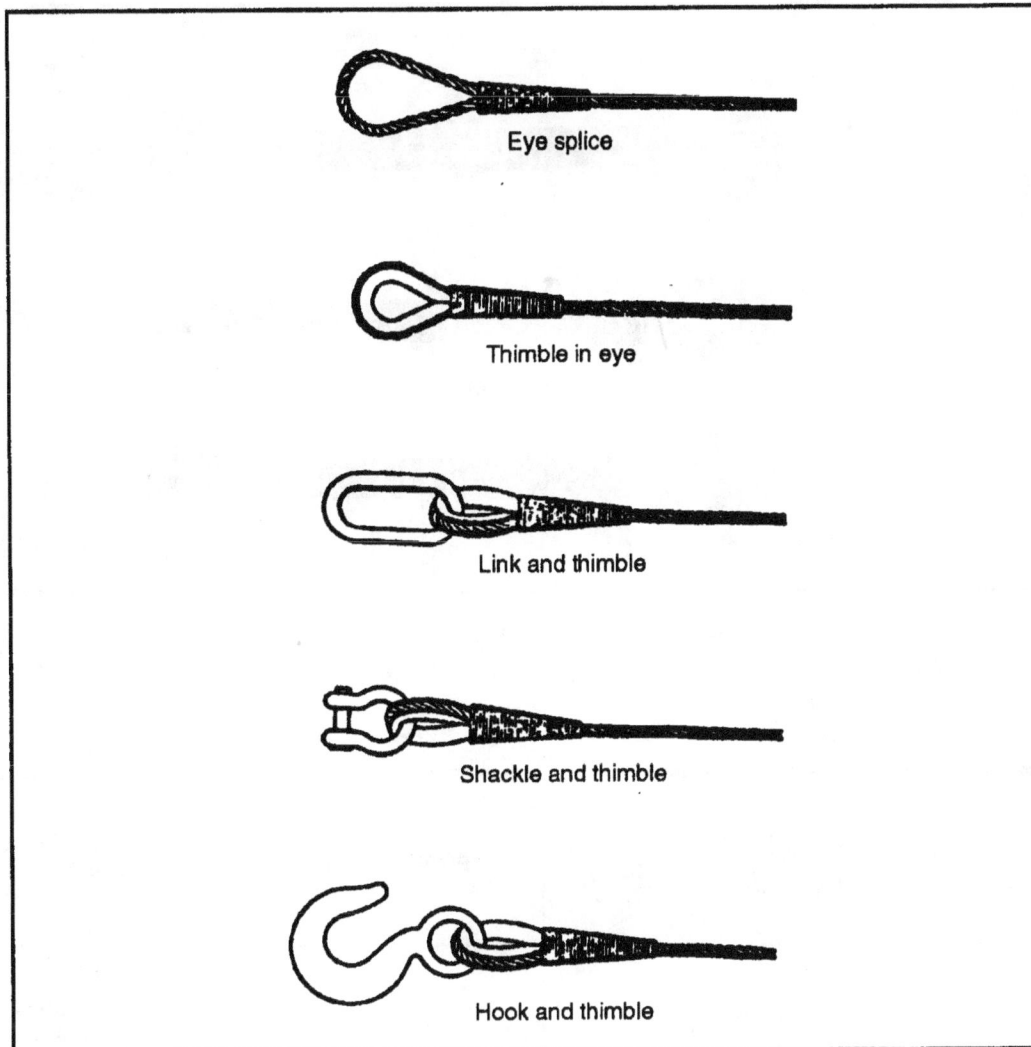

Figure 2-49. Attachments used with eye splice

END FITTINGS

An end fitting may be placed directly on wire rope. Fittings that are easily and quickly changed are clips, clamps, and wedge sockets.

CLIPS

Wire-rope clips are reliable and durable (see *Figure 2-50, page 2-36*) Use them, repeatedly, to make eyes in wire rope, either for a simple eye or an eye reinforced with a thimble, or to secure a wire-rope line or anchorage to another wire rope. *Table 2-3, page 2-36* shows the number and spacing of clips and the proper torque to apply to the nuts of the clips. After installing all the clips, tighten the clip farthest from the eye (thimble) with a torque wrench. Next, place the rope under tension and tighten the clip next to the clip you tightened first. Tighten the remaining clips in order, moving toward the loop (thimble). After placing the rope in service, tighten the clips again immediately

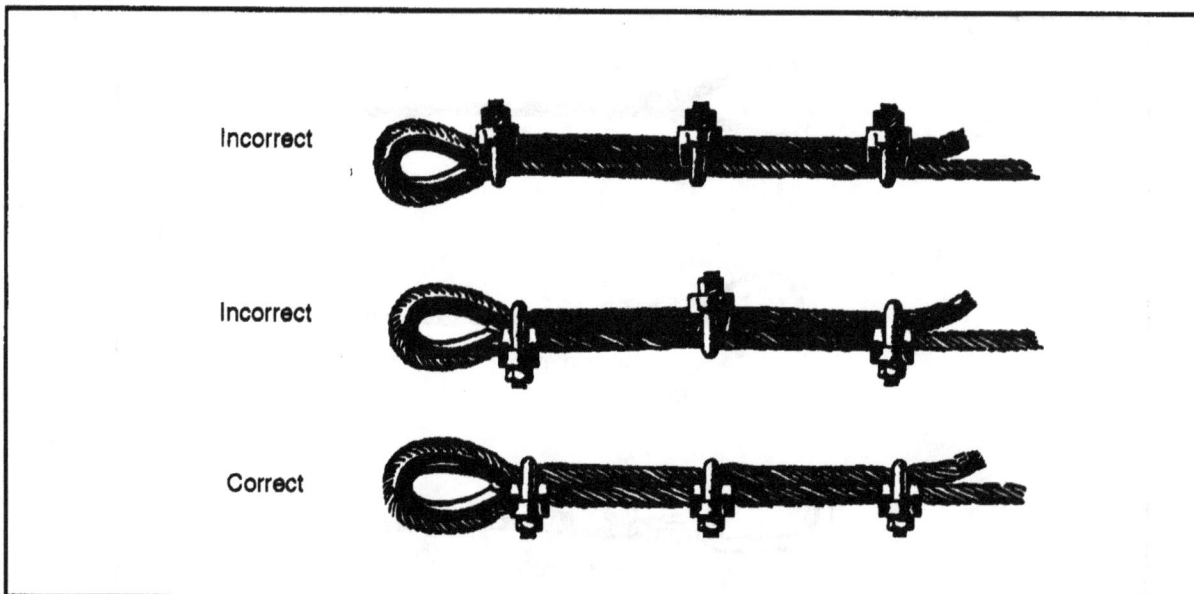

Figure 2-50. Wire-rope clips

Table 2-3. Assembling wire-rope eye-loop connections

Wire-Rope Diameter		Nominal Size of Clips	Number of Clips	Spacing of Clips		Torque to be Applied to Nuts of Clips	
(Inches)	(mm)	(Inches)		(Inches)	(mm)	(ft–lb)	(m–kg x 0.1382)
5/16	(7.95)	3/8	3	2	(50)	25	(3.5)
3/8	(9.52)	3/8	3	2 1/4	(57)	25	(3.5)
7/16	(11.11)	1/2	4	2 3/4	(70)	40	(5.5)
1/2	(12.70)	1/2	4	3	(76)	40	(5.5)
5/8	(15.85)	5/8	4	3 3/4	(95)	65	(9.0)
3/4	(19.05)	3/4	4	4 1/2	(114)	100	(14.0)
7/8	(22.22)	1	5	5 1/4	(133)	165	(23.0)
1	(25.40)	1	5	6	(152)	165	(23.0)
1 1/4	(31.75)	1 1/4	5	7 1/2	(190)	250	(35.0)
1 3/8	(34.92)	1 1/2	6	8 1/4	(210)	375	(52.0)
1 1/2	(38.10)	1 1/2	6	9	(230)	375	(52.0)
1 3/4	(44.45)	1 3/4	6	101/2	(267)	560	(78.0)

NOTE: The spacing of clips should be six times the diameter of the wire rope. To assemble an end-to-end connection, the number of clips indicated above should be increased by two, and the proper torque indicated above should be used on all clips. U-bolts are reversed at the center of connection so that the U-bolts are on the dead (reduced load) end of each wire rope.

after applying the working load and at frequent intervals thereafter. Retightening is necessary to compensate for the decrease in rope diameter that occurs when the strands adjust to the lengthwise strain caused by the load. Position the clips so that they are immediately accessible for inspection and maintenance.

CLAMPS

A wire clamp can be used with or without a thimble to make an eye in wire rope (see *Figure 2-51*). Ordinarily, use a clamp to make an eye without a thimble. It has about 90 percent of the strength of the rope. Tighten the two end collars with wrenches to force the clamp to a good snug fit. This crushes the pieces of rope firmly against each other.

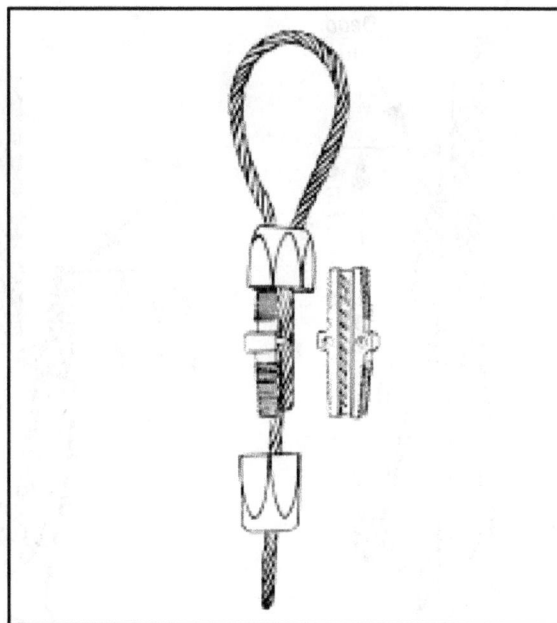

Figure 2-51. Wire-rope clamps

WEDGE SOCKET

Use a wedge-socket end fitting when it is necessary to change the fitting at frequent intervals (see *Figure 2-52*, *page 2-38*). The efficiency is about two-thirds of the strength of the rope. It is made in two parts. The socket itself has a tapered opening for the wire rope and a small wedge to go into this tapered socket. The loop of wire rope must be inserted in the wedge socket so that the standing part of the wire rope will form a nearly direct line to the clevis pin of the fitting. A properly installed wedge-socket connection will tighten when a strain is placed on the wire rope.

BASKET-SOCKET END FITTING

The basket-socket end fittings include closed sockets, open sockets, and bridge sockets (see Figure 2-53, page 2-38). This socket is ordinarily attached to the end of the rope with molten zinc and is a permanent end rifting. If this fitting is properly made up, it is as strong as the rope itself. In all cases, the wire rope should lead from the socket in line with the axis of the socket.

POURED METHOD

The poured basket socket is the most satisfactory method in use (see *Figure 2-54, page 2-39*). If the socketing is properly done, a wire rope, when tested to destruction, will break before it will pull out from the socket.

> **WARNING**
> *Never use babbitt, lead, or dry method to attach a basket socket end fitting.*

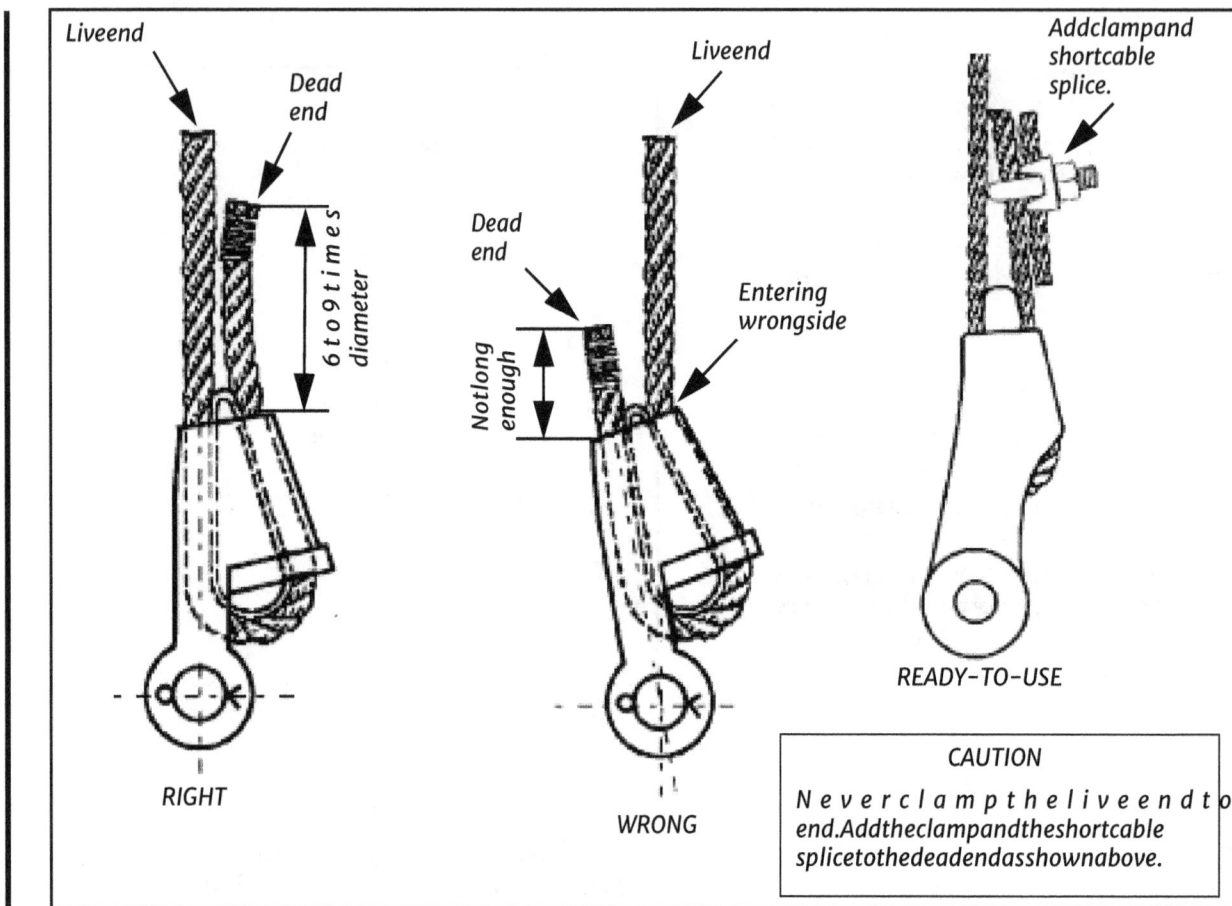

Liveend

Dead end

6 to 9 times diameter

RIGHT

Liveend

Dead end

Not long enough

Entering wrongside

WRONG

Add clamp and short cable splice.

READY-TO-USE

CAUTION

Never clamp the live end to the dead end. Add the clamp and the short cable splice to the dead end as shown above.

Figure 2-52. Wedge socket

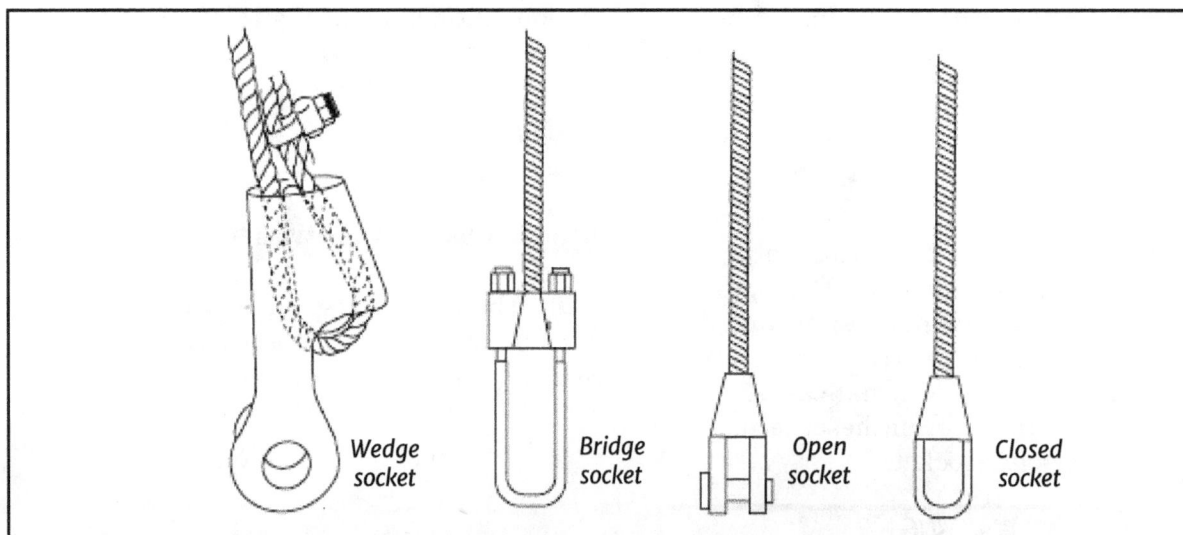

Wedge socket

Bridge socket

Open socket

Closed socket

Figure 2-53. Basket-socket end fittings

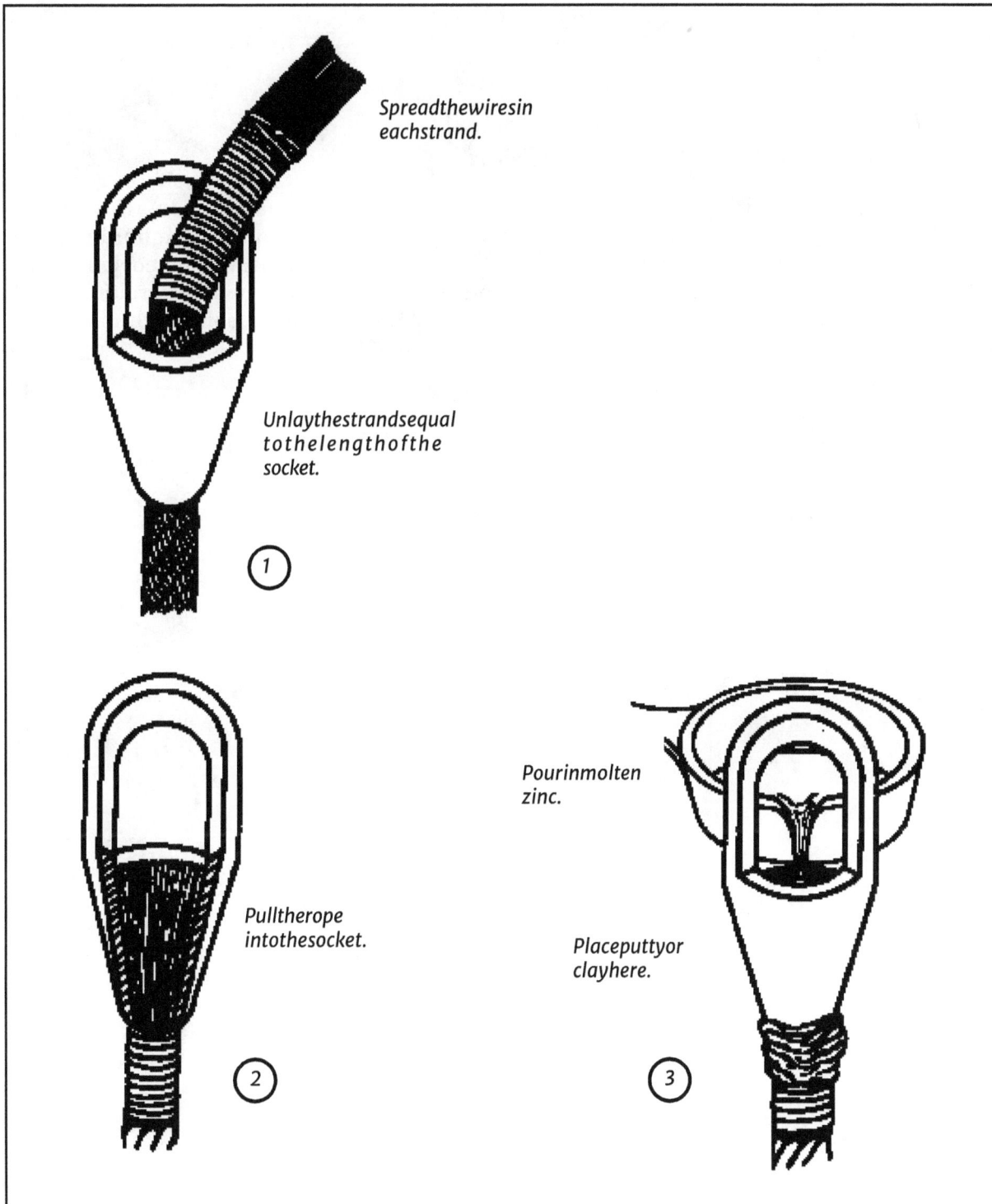

Spread the wires in each strand.

Unlay the strands equal to the length of the socket.

①

Pull the rope into the socket.

②

Pour in molten zinc.

Place putty or clay here.

③

Figure 2-54. Attaching basket sockets by pouring

STANCHIONS

The standard pipe stanchion is made up of a 1-inch diameter pipe (see *Figure 2-56*). Each stanchion is 40 inches long. Two 3/4-inch wire-rope clips are fastened through holes in the pipe with the centers of the clips 36 inches apart. Use this stanchion, without modifying it, for a suspended walkway that uses two wire ropes on each side. However, for handlines, remove or leave off the lower wire-rope clip. For more information on types and uses of stanchions, see *TM 5-270*.

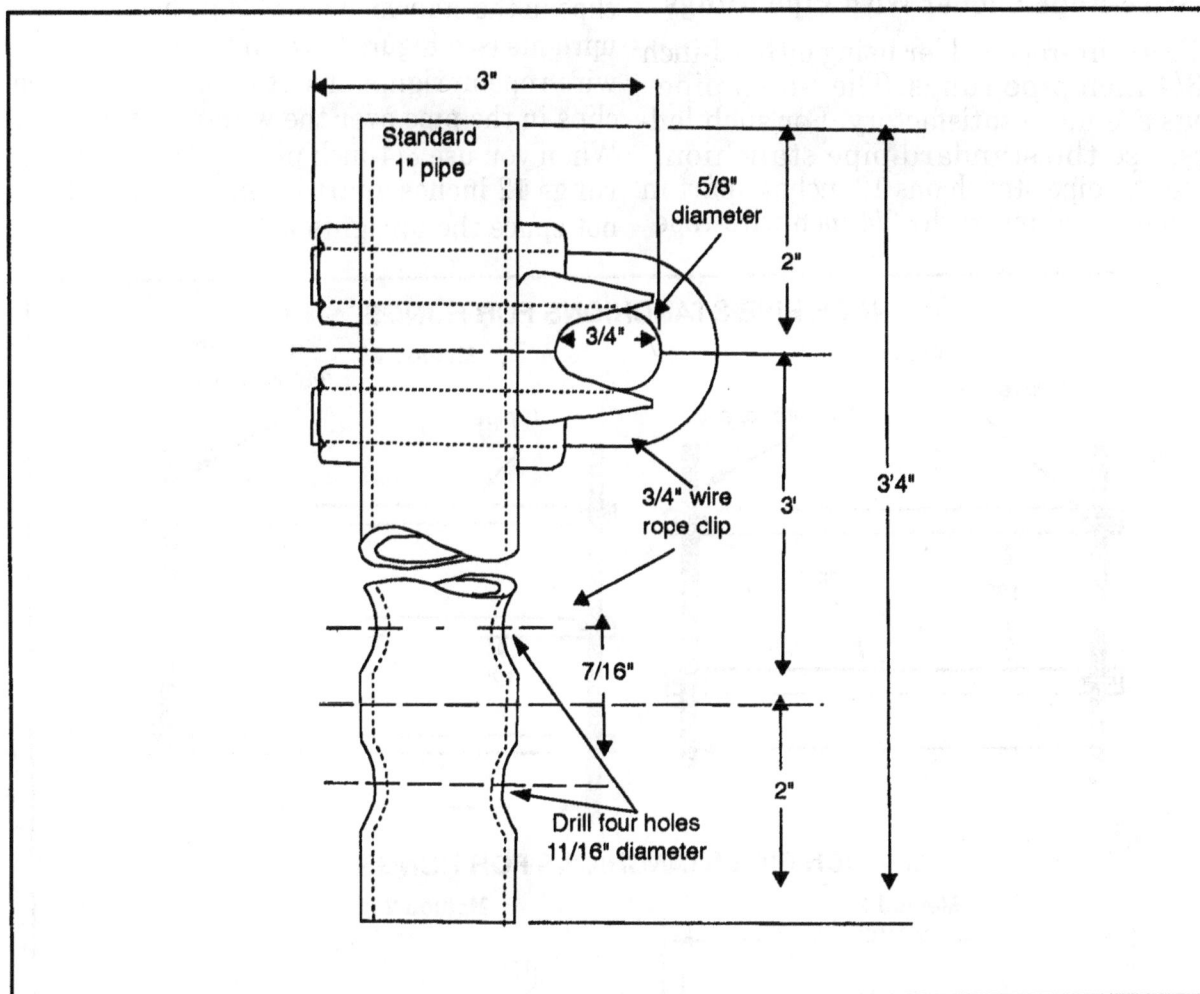

Figure 2-56. Iron-pipe stanchions

Section IV. Rope Ladders

Ropes may be used in the construction of hanging ladders and standoff ladders.

HANGING LADDERS

Hanging ladders are made of wire or fiber rope anchored at the top and suspended vertically. They are difficult to ascend and descend, particularly for a man carrying a pack or load and should be used only when necessary. The uprights of hanging ladders may be made of wire or fiber rope and anchored at the top and bottom.

WIRE-ROPE LADDERS

Wire-rope uprights with pipe rungs make the most satisfactory hanging ladders because they are more rigid and do not sag as much as hanging ladders made of other material. Wire-rope uprights with wire-rope rungs are usable.

Wire-Rope Ladder With Pipe Rungs

Make a wire-rope ladder using either l-inch or 3/4-inch pipe rungs. The l-inch pipe rungs are more satisfactory. For such ladders, use the standard pipe stanchion. Space the pipe stanchions 12 inches apart in the ladder and insert the 3/4-inch wire-rope clips in the stanchion over 3/4-inch wire-rope uprights (see Figure 2-57). If you use 3/8-inch wire-rope uprights, insert 3/8-inch wire-rope clips in the pipe over the wire-rope uprights. When you use 3/4-inch pipe rungs, space the rungs 12 inches apart in the ladder, but do not space the uprights more than 12 inches

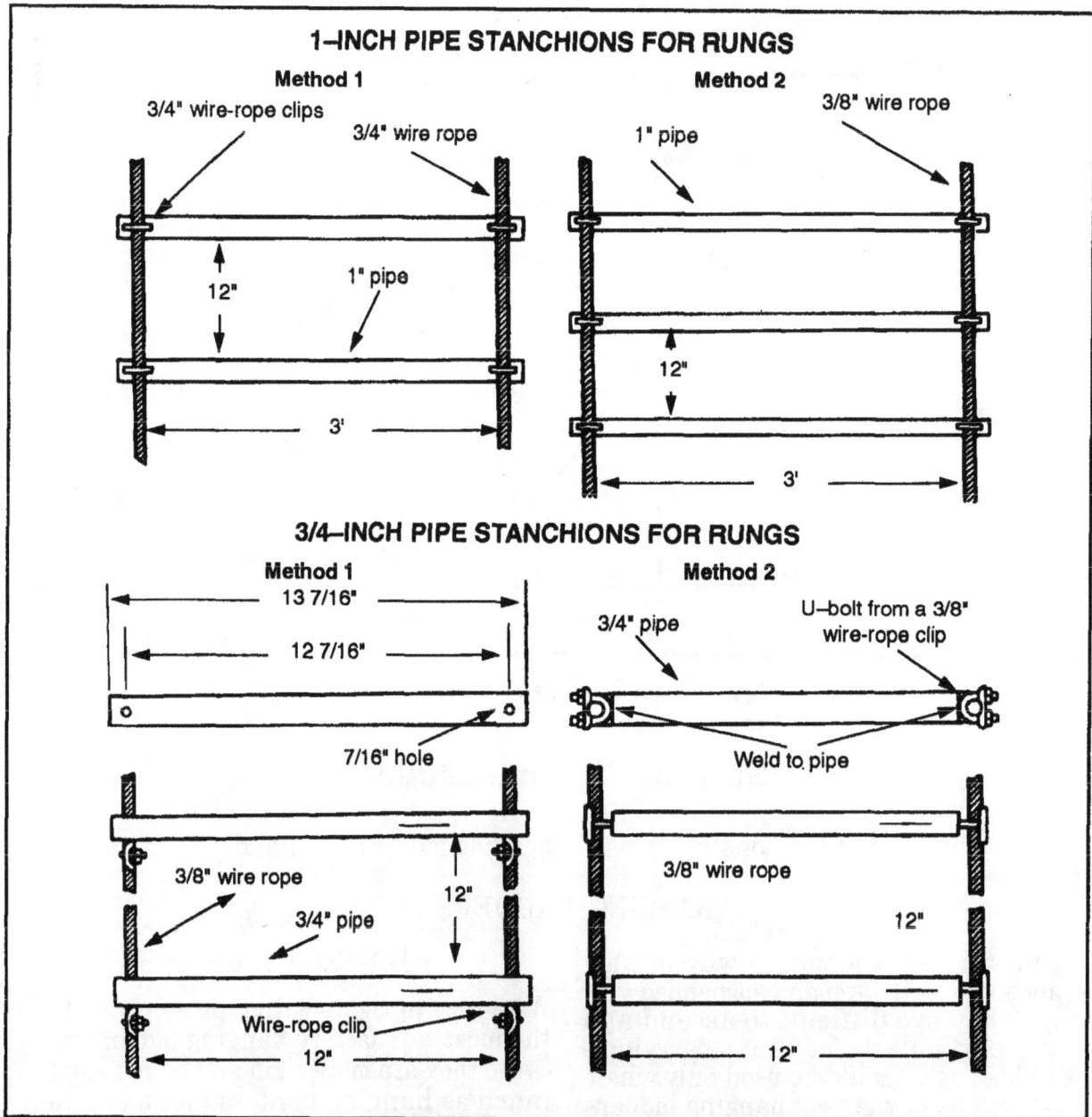

Figure 2-57. Pipe rungs

apart because of using weaker pipe. The rungs may be fastened in place by two different methods. In one method, drill a 7/16-inch diameter hole at each end of each pipe rung and thread 3/8-inch wire-rope uprights through the holes. To hold each rung in place, fasten a 3/8-inch wire-rope clip about the wire-rope upright at each end of each rung after the rung is in its final position. In the other method, cut the pipe rungs 12 inches long and weld the U-bolt of a 3/8-inch rope clip to each end. Space the rungs 12 inches apart on the 3/8-inch wire-rope uprights. Place the saddle of the wire-rope clips and the nuts on the U-bolts; tighten the nuts to hold the rungs in place.

Wire-Rope Ladder With Wire-Rope Rungs

Make a wire-rope ladder with wire-rope rungs by laying the 3/8-inch diameter wire-rope uprights on the ground. Lay out the first length in a series of U-shaped bends. Lay out the second length in a similar manner with the U-shaped bends in the opposite direction from those in the first series and the horizontal rung portions overlapping (see *Figure 2-58*)Fasten a 3/8-inch wire-rope clip on the overlapping rung portions at each end of each rung to hold them firm.

FIBER-ROPE LADDERS

Fiber-rope uprights with wood or fiber-rope rungs are difficult to use because their greater flexibility causes them to twist when they are being used. Place a log at the break of the ladder at the top to hold the uprights and rungs away from a rock face to provide better handholds and footholds. A single rock anchor at the bottom of the ladder is usually sufficient. You can also use a pile of rocks as the bottom anchor for fiber-rope hanging ladders.

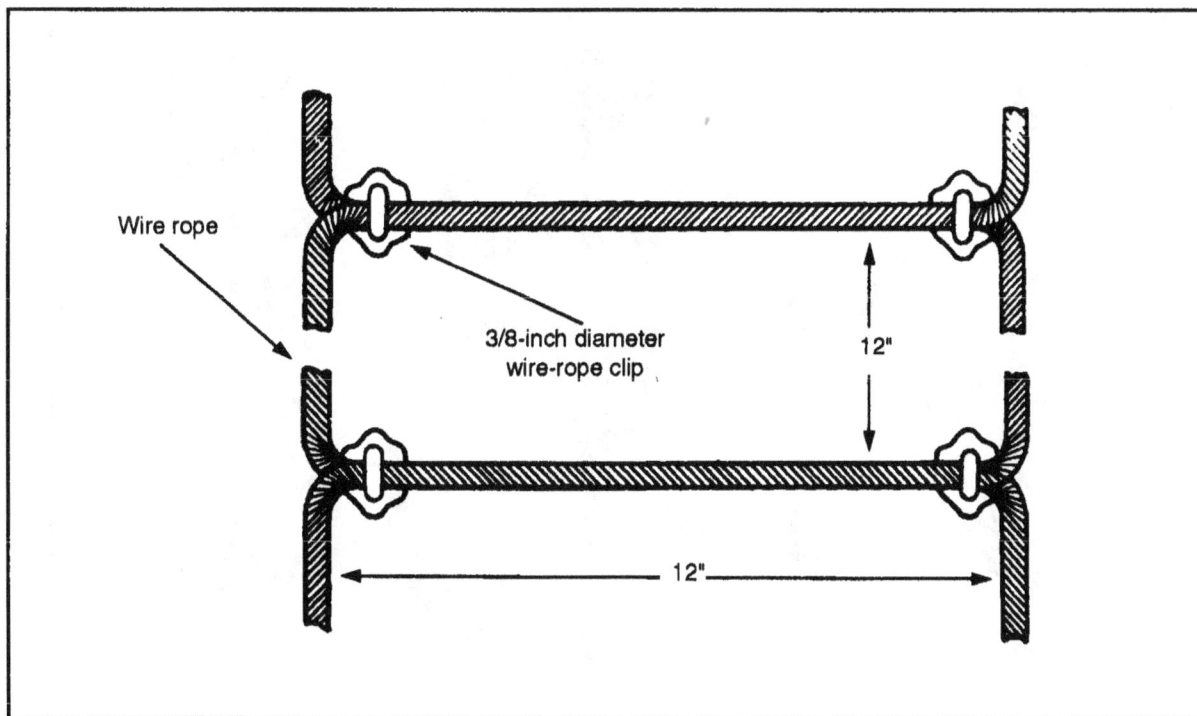

Figure 2-58. Wire-rope rungs

Fiber-Rope Ladder With Fiber-Rope Rungs

Make fiber-rope ladders with fiber-rope rungs by using two or three uprights. When you use three uprights, make a loop in the center upright at the position of each rung (see Figure 2-59). Space the two outside uprights 20 inches apart. A loop and a single splice hold each end of each rung to the outside upright. A loop in the center of the rung passes through the loop in the center upright. If you use only two uprights, hold the rungs in place by a loop and a rolling hitch or a single splice at each upright. The two uprights must be closer together, with shorter rungs, to stiffen the ladder. Ladders of either type are very flexible and difficult to climb.

Fiber-Rope Ladder With Wood Rungs

Make fiber-rope ladders with wood rungs by using finished lumber or native material for rungs (see *Figure 2-60*). When you use native material, cut the rungs from 2-inch-diameter material about 15 inches long. Notch the ends of each rung and fasten the rung to the fiber-rope upright with a clove hitch. Space the rungs 12 inches apart. Twist a piece of seizing wire about the back of the clove hitch to make it more secure and in a manner that will not snag the clothing of persons climbing the ladder. If you make the rungs of finished lumber, cut them to size and drill a 3/4-inch hole at each end. Oak lumber is best for this purpose. Put a 1/4-inch by 2-inch carriage bolt horizontally through each end near the vertical hole to prevent splitting. Tie an overhand knot in the upright to support the rung. Then thread the upright through the 3/4-inch hole in the rung. Tie a second overhand knot in the upright before you thread it through the next rung. Continue this Procedure until you reach the desired length of the ladder.

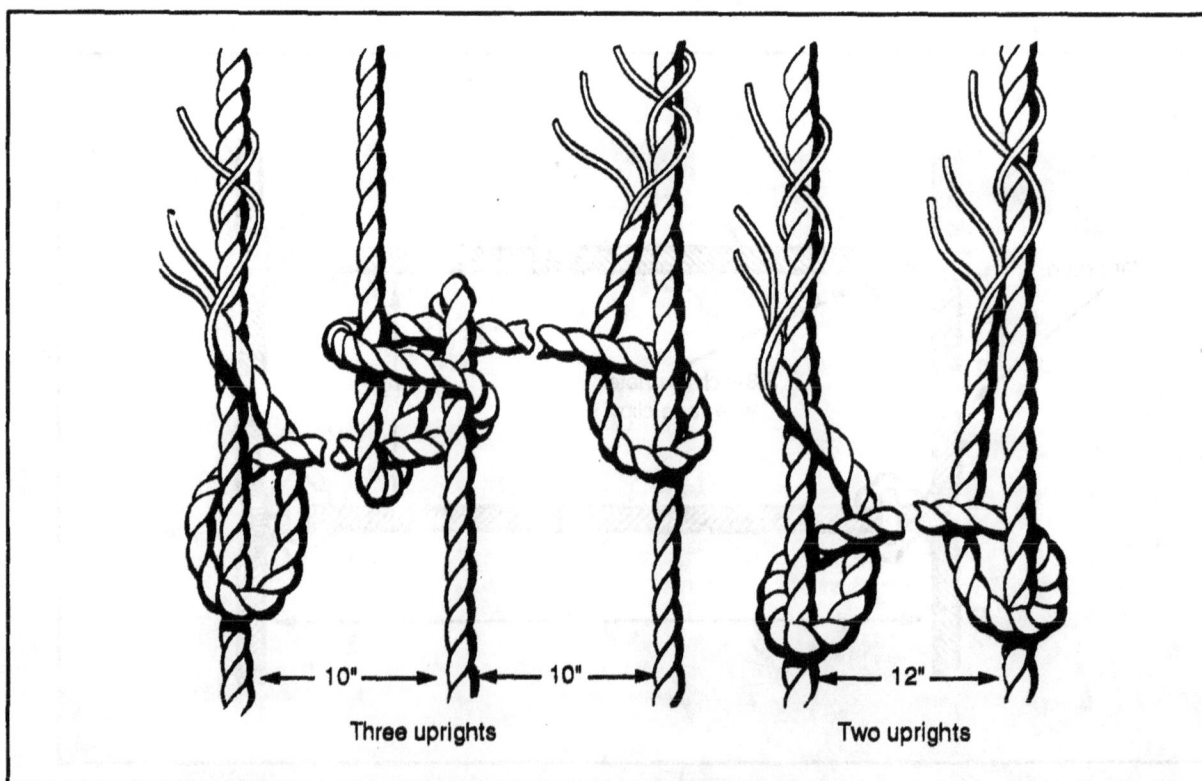

Figure 2-59. Fiber-rope rungs

STANDOFF LADDERS

Standoff ladders are easier to climb than hanging ladders because they have two wood or metal uprights that hold them rigid, and they are placed at an angle. Both types of ladders can be prefabricated and transported easily. One or two standoff ladders are adequate for most purposes, but three or four hanging ladders must be provialed for the same purpose because they are more difficult to use.

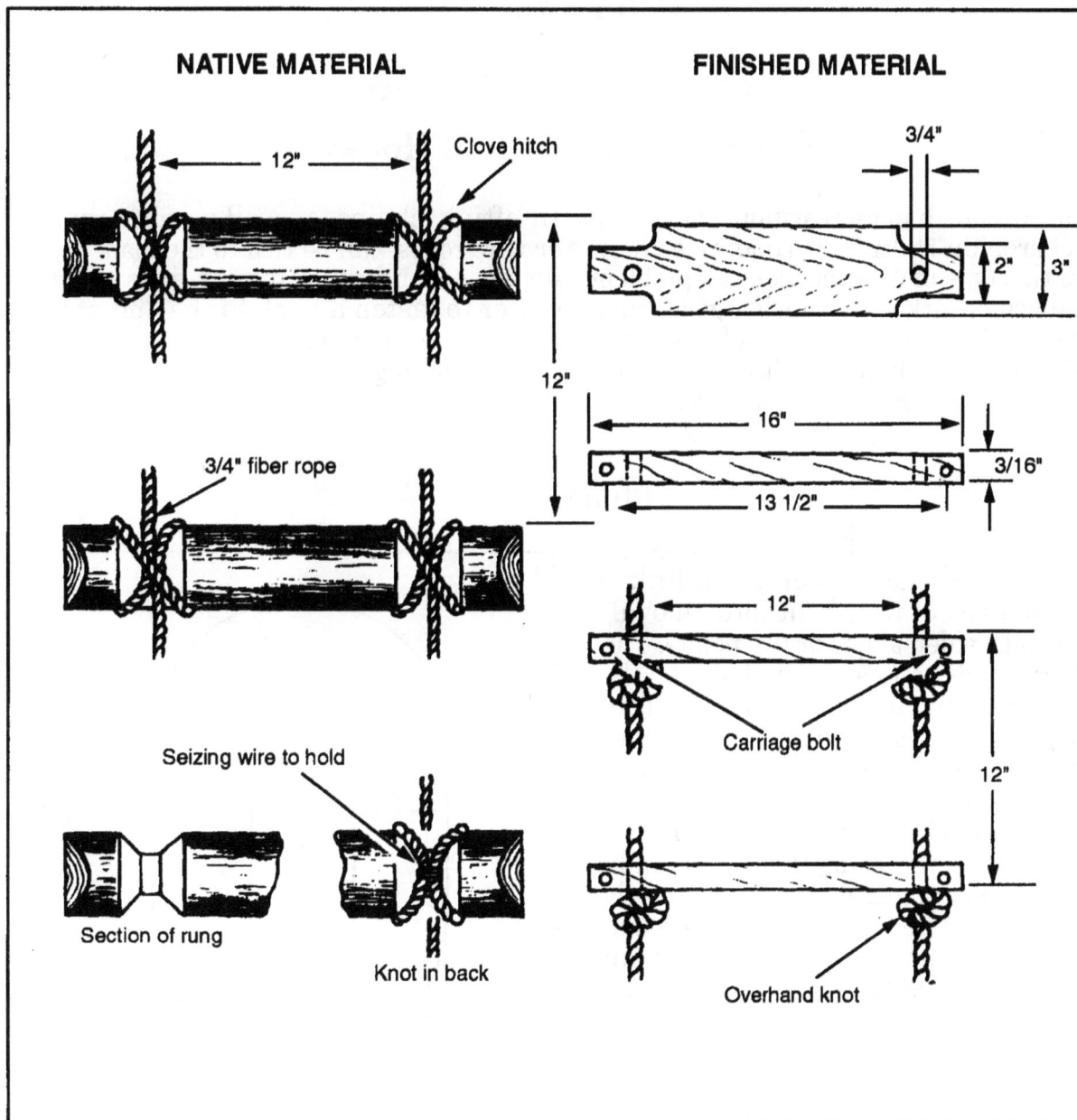

NATIVE MATERIAL

Clove hitch

12"

3/4" fiber rope

12"

Seizing wire to hold

Section of rung

Knot in back

FINISHED MATERIAL

3/4"

2" 3"

16"

3/16"

13 1/2"

12"

Carriage bolt

12"

Overhand knot

Figure 2-60. Wood rungs

CHAPTER *3*

Hoists

Section I. Chains and Hooks

Chains are much more resistant to abrasion and corrosion than wire rope; use them where this type of deterioration is a problem, as in marine work where anchor gear must withstand the corrosive effects of seawater. You can also use chains to lift heavy objects with sharp edges that would cut wire.

In lifting, chains, as well as fiber ropes or wire ropes, can be tied to the load. But for speed and convenience, it is much better to fasten a hook to the end of the lifting line. Also, you can use hooks are in constructing blocks.

CHAINS

Chains are made up of a series of links fastened through each other. Each link is made of a rod of wire bent into an oval shape and welded at one or two points. The weld ordinarily causes a slight bulge on the side or end of the link (see *Figure 3-1).*The chain size refers to the diameter, in inches, of the rod used to make the link. Chains usually stretch under excessive loading so that the individual links bend slightly. Bent links are a warning that the chain has been overloaded and might fail suddenly under a load. Wire rope, on the other hand, fails a strand at a time, giving warning before complete failure occurs. If a chain is equipped with the proper hook, the hook should start to fail first, indicating that the chain is overloaded.

Several grades and types of chains are available.

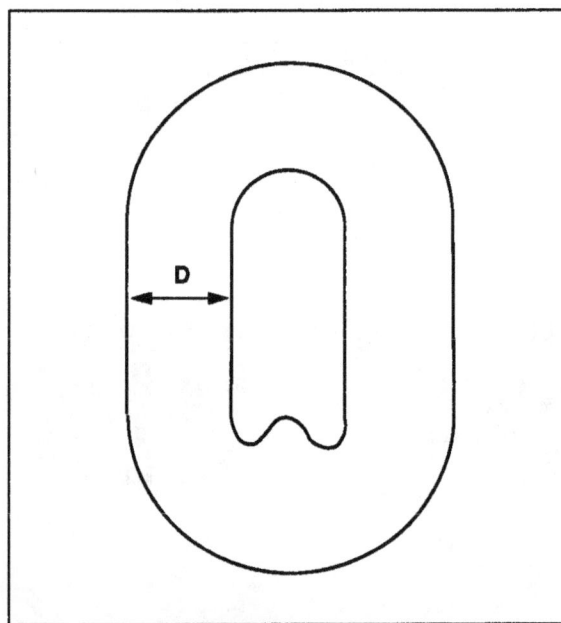

Figure 3-1. Link thickness

STRENGTH OF CHAINS

To determine the SWC on a chain, apply a FS to the breaking strength. The SWC ordinarily is assumed to be about one-sixth of the BS, giving a FS of 6. Table 3-1 lists SWC for various chains. You can approximate the SWC of an open-link chain by using the following rule of thumb:

$$SWC = 8D^2$$

SWC = Safe working capacity, in tons

D = Smallest link thickness or least diameter measured in inches (see Figure 3-1, page 3-1)

Example: Using the rule of thumb, the SWC of a chain with a link thickness of 3/4 inch is—

$$SWC = 8D^2 = 8 (3/4)^2 = 4.5 \text{ tons or } 9,000 \text{ pounds}$$

The figures given assume that the load is applied in a straight pull rather than by an impact. An impact load occurs when an object is dropped suddenly for a distance and stopped. The impact load in such a case is several times the weight of the load.

CARE OF CHAINS

When hoisting heavy metal objects using chains for slings, insert padding around the sharp corners of the load to protect the

Table 3-1. Properties of chains (FS 6)

Size*	Approximate Weight per Linear Foot (pounds)	SWC (pounds)			
		Common Iron	High-Grade Iron	Soft Steel	Special Steel
1/4	0.8	512	563	619	1,240
3/8	1.7	1,350	1,490	1,650	3,200
1/2	2.5	2,250	2,480	2,630	5,250
5/8	4.3	3,470	3,810	4,230	7,600
3/4	5.8	5,070	5,580	6,000	10,500
7/8	8.0	7,000	7,700	8,250	14,330
1	10.7	9,300	10,230	10,600	18,200
11/8	12.5	9,871	10,858	11,944	21,500
11/4	16.0	12,186	13,304	14,634	26,300
13/8	18.3	14,717	16,188	17,807	32,051

*Size listed is the diameter, in inches, of one side of a link.

chain links from being cut. The padding may be either planks or heavy fabric. Do not permit chains to twist or kink when under strain. Never fasten chain links chain together with bolts or wire because such connections weaken the chain and limit its SWC. Cut worn or damaged links out of the chain and replace them with a cold-shut link. Close the cold-shut link and weld it to equal the strength of the other links.

Cut the smaller chain links with a bolt cutter; cut large chain links with a hacksaw or an oxyacetylene torch. Inspect the chains frequently, depending on the amount of use. Do not paint chains to prevent rusting because the paint will interfere with the action of the links. Instead, apply a light coat of lubricant and store them in a dry and well-ventilated place.

HOOKS

The two general types of hooks available are the slip hook and the grab hook (see *Figure 3-2*)Slip hooks are made so that the inside curve of the hook is an arc of a circle and may be used with wire rope. chains. or fiber rope. Chain links can slip through a slip hook so the loop formed in the chain will tighten under a load. Grab hooks have an inside curve that is nearly U-shaped so that the hook will slip over a link of chain edgeways but will not permit the next link to slip through. Grab hooks have a more limited range of use than slip hooks. They are used on chains when the loop formed with the hook is not intended to close up around the load.

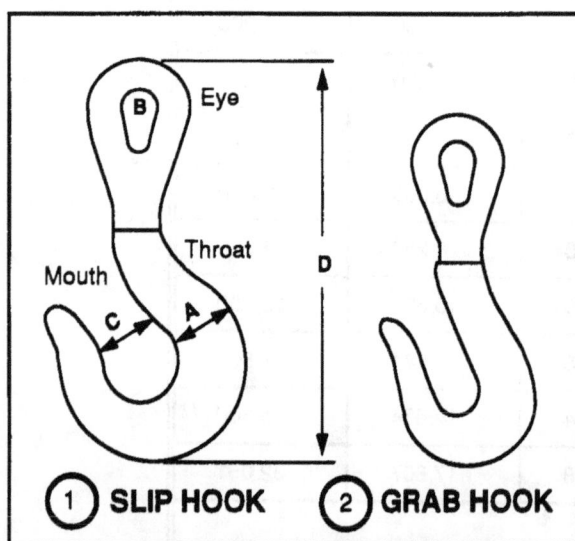

Figure 3-2. Types of hooks

STRENGTH OF HOOKS

Hooks usually fail by straightening. Any deviation from the original inner arc indicates that the hook has been overloaded. Since you can easily detect evidence of overloading the hook. you should use a hook that is weaker than the chain to which it is attached. With this system. hook distortion will occur before the chain is overloaded. Discard severely distorted. cracked. or badly worn hooks because they are dangerous. *Table 3-2. page 3-4* lists SWCs on hooks. Approximate the SWC of a hook by using the following rule of thumb:

$$SWC = D^2$$

D = *the diameter in inches of the hook where the inside of the hook starts its arc (see Figure 3-3. page 3-5)*

Thus. the SWC of a hook with a diameter of 1 1/4 inches is as follows:

$$SWC = D^2 = (1\ 1/4)^2\ 16\ tons\ or\ 3,125\ pounds$$

MOUSING OF HOOKS

In general. always "mouse" a hook as a safety measure to prevent slings or ropes from jumping off. To mouse a hook after the sling is on the hook. wrap the wire or heavy twine 8 or 10 turns around the two sides of the hook (see*Figure 3-4. page 3-5).*

Table 3-2. Safe loads on hooks

Diameter of Metal A* (inches)	Inside Diameter of Eye B (Inches)	Width of Opening C (Inches)	Length of Hook D (Inches)	SWC of Hooks, (pounds)
11/16	1/8	1 1/16	4 15/16	1,200
3/4	1	1 1/3	5 13/32	1,400
7/8	1 1/8	1 1/4	6 1/4	2,400
1	1 1/4	1 3/8	6 7/8	3,400
1 1/8	1 3/8	1 1/2	7 5/8	4,200
1 1/4	1 1/2	1 11/16	8 19/32	5,000
1 3/8	1 5/8	1 7/8	9 1/2	6,000
1 1/2	1 3/4	2 1/16	10 11/32	8,000
1 5/8	2	2 1/4	11 21/32	9,400
1 7/8	2 3/8	2 1/2	13 9/32	11,000
2 1/4	2 3/4	3	14 13/16	13,600
2 5/8	3 1/8	3 3/8	16 1/2	17,000
3	3 1/2	4	19 3/4	24,000

*For reference to A, B, C, or D, see Figure 3-2.

Complete the process by winding several turns of the wire or twine around the sides of the mousing and tying the ends securely. Mousing also helps prevent straightening of the hook but does not strengthen it materially.

INSPECTING CHAINS AND HOOKS

Inspect chains, including the hooks, at least once a month; inspect those that are used for heavy and continuous loading more frequently. Give particular attention to the small radius fillets at the neck of hooks for any deviation from the original inner arc. Examine each link and hook for small dents and cracks, sharp nicks or cuts, worn surfaces, and distortions. Replace those that show any of these weaknesses. If several links are stretched or distorted, do not use the chain; it probably was overloaded or hooked improperly, which weakened the entire chain.

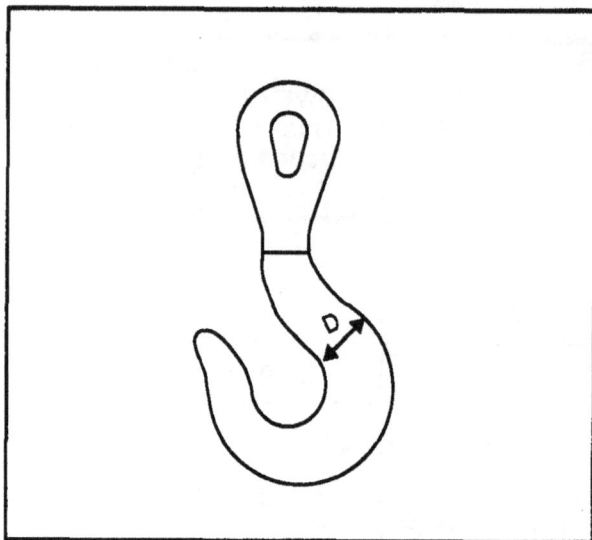

Figure 3-3. Hook thickness (diameter)

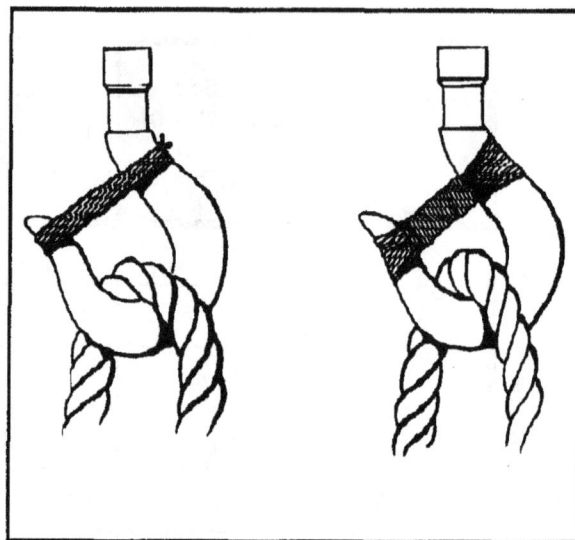

Figure 3-4. Mousing hooks

Section II. Slings

The term "sling" includes a wide variety of designs. Slings may be made of fiber rope, wire rope, or chain.

Fiber rope makes good slings because of its flexibility, but it is more easily damaged by sharp edges on the material hoisted than are wire rope or chain slings. Fiber-rope slings are used for lifting comparatively light loads and for temporary jobs.

Wire rope is widely used for slings because it has a combination of strength and flexibility. Properly designed and appropriately fabricated wire-rope slings are the safest type of slings. They do not wear away as do slings made of fiber rope, nor do they lose their strength from exposure as rapidly. They also are not susceptible to the "weakest link" condition of chains caused by the uncertainty of the strengths of the welds. The appearance of broken wires clearly indicates the fatigue of the metal and the end of the usefulness of the sling.

Chain slings are used especially where sharp edges of metal would cut wire rope or where very hot items are lifted, as in foundries or blacksmith shops.

Barrel slings can be made with fiber rope to hold barrels horizontally or vertically.

TYPES OF SLINGS

The sling for lifting a given load may be—

- An endless sling.
- A single sling.
- A combination sling (several single slings used together).

Each type or combination has its particular advantages that must be considered when selecting a sling for a given purpose.

ENDLESS SLINGS

The endless sling is made by splicing the ends of a piece of wire rope or fiber rope together or by inserting a cold-shut link in a chain. Cold-shut links should be welded after insertion in the chain. These endless slings are simple to handle and may be used in several different ways to lift loads (see *Figure 3-5, page 3-6*).

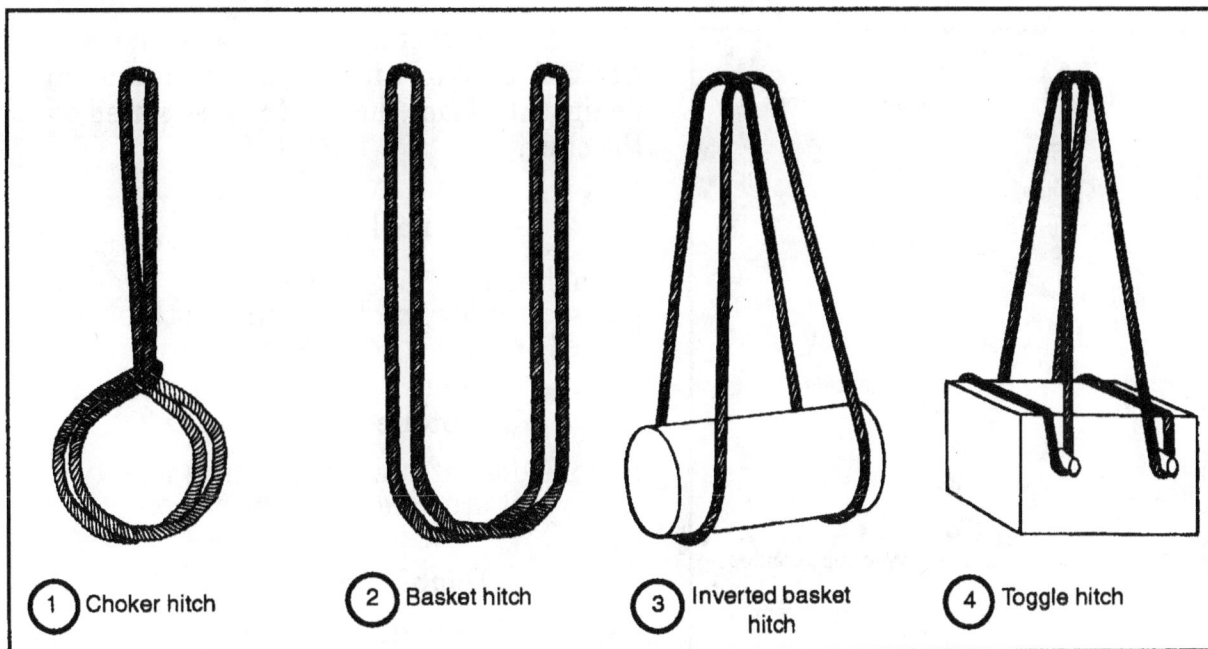

Figure 3-5. Endless slings

Choker or Anchor Hitch

A common method of using an endless sling is to cast the sling under the load to be lifted and insert one loop through the other and over the hoisting hook. When the hoisting hook is raised, one side of the choker hitch is forced down against the load by the strain on the other side, forming a tight grip on the load.

Basket Hitch

With this hitch, the endless sling is passed around the object to be lifted and both remaining loops are slipped over the hook.

Inverted Basket Hitch

This hitch is very much like the simple basket hitch except that the two parts of the sling going under the load are spread wide apart.

Toggle Hitch

The toggle hitch is used only for special applications. It is actually a modification of the inverted basket hitch except that the line passes around toggles fastened to the load rather than going around the load itself.

SINGLE SLINGS

A single sling can be made of wire rope, fiber rope, or chain. Each end of a single sling is made into an eye or has an attached hook (see *Figure 3-6*).In some instances, the ends of a wire rope are spliced into the eyes that are around the thimbles, and one eye is fastened to a hook with a shackle. With this type of single sling, you can remove the shackle and hook when desired. You can use a single sling in several different ways for hoisting (see *Figure 3-6*).It is advisable to have four single slings of wire rope available at all times. These can be used singly or in combination, as necessary.

Choker or Anchor Hitch

A choker or anchor hitch is a single sling that is used for hoisting by passing one eye

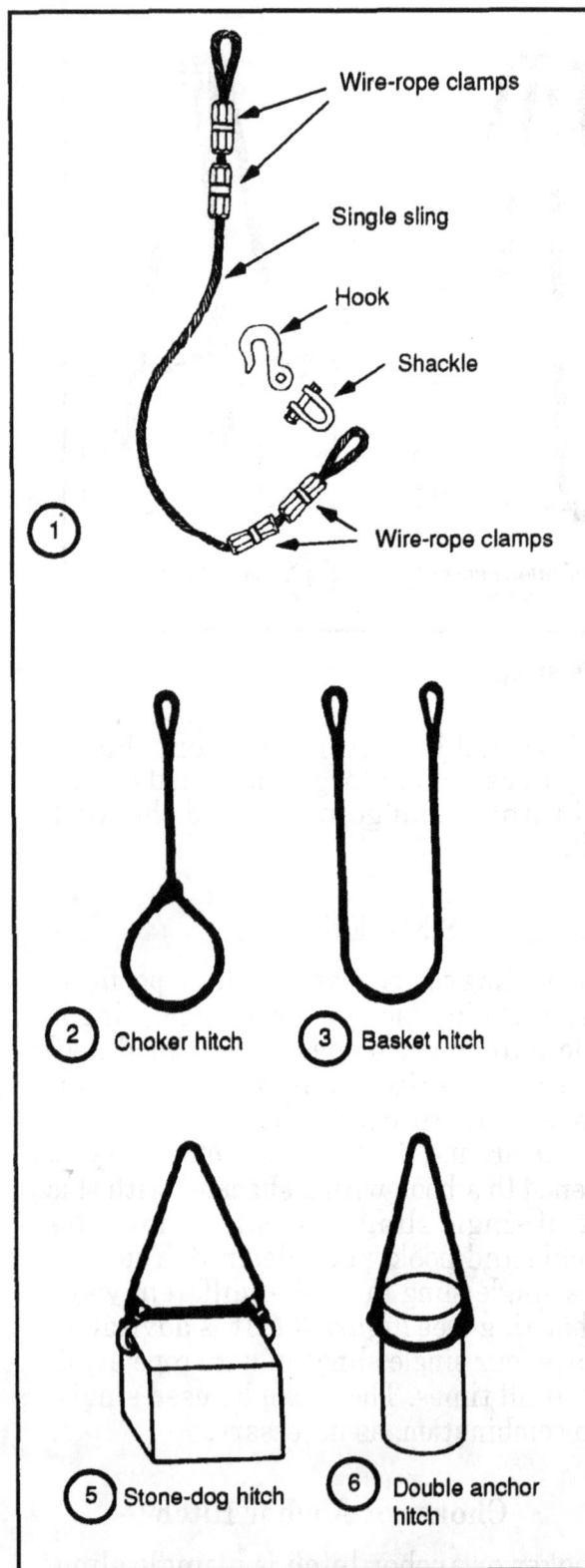

Figure 3-6. Single slings

through the other eye and over the hoisting hook. A choker hitch will tighten down against the load when a strain is placed on the sling.

Basket Hitch

A basket hitch is a single sling that is passed under the load with both ends hooked over the hoisting hook.

Stone-Dog Hitch

A stone-dog hitch is single slings with two hooks that are used for lifting stone.

Double Anchor Hitch

This hitch is used for hoisting drums or other cylindrical objects where it is necessary for the sling to tighten itself under strain and lift by friction against the sides of the cylinder.

COMBINATION SLINGS

Single slings can be combined into bridle slings, basket slings, and choker slings to lift virtually any type of load. Either two or four single slings can be used in a given combination. Where greater length is required, two of the single slings can be combined into a longer single sling. One of the problems in lifting heavy loads is in fastening the bottom of the sling legs to the load in such a way that the load will not be damaged. Lifting eyes are fastened to many pieces of equipment at the time it is manufactured. On large crates or boxes, the sling legs may be passed under the object to form a gasket sling. A hook can be fastened to the eye on one end of each sling leg to permit easier fastening on some loads. Where the load being lifted is heavy enough or awkward enough, a four-leg sling may be required. If a still greater length of sling is required, two additional slings can be used in conjunction with the four-leg sling to form a double basket.

PALLETS

A problem in hoisting and moving loads sometimes occurs when the items to be lifted are packaged in small boxes and the individual boxes are not crated. In this case, it is entirely too slow to pick up each small box and move it separately. Pallets, used in combination with slings, provide an efficient method of handling such loads. The pallets can be made up readily on the job out of 2- by 8-inch timbers that are 6 or 8 feet long and are nailed to three or four heavy cross members, such as 4- by 8-inch timbers. Several pallets should be made up so that one pallet can be loaded while the pallet previously loaded is being hoisted. As each pallet is unloaded, the next return trip of the hoist takes the empty pallet back for loading.

SPREADERS

Occasionally, it is necessary to hoist loads that are not protected sufficiently to prevent crushing by the sling legs. In such cases, spreaders may be used with the slings (see *Figure* 3-7). Spreaders are short bars or pipes with eyes on each end. The sling leg passes through the eye down to its connection with the load. By setting spreaders in the sling legs above the top of the load, the angle of the sling leg is changed so that crushing of the load is prevented. Changing the angle of the sling leg may increase the stress in that portion of the sling leg above the spreaders. The determining factor in computing the safe lifting capacity of the sling is the stress (or tension) in the sling leg above the spreader.

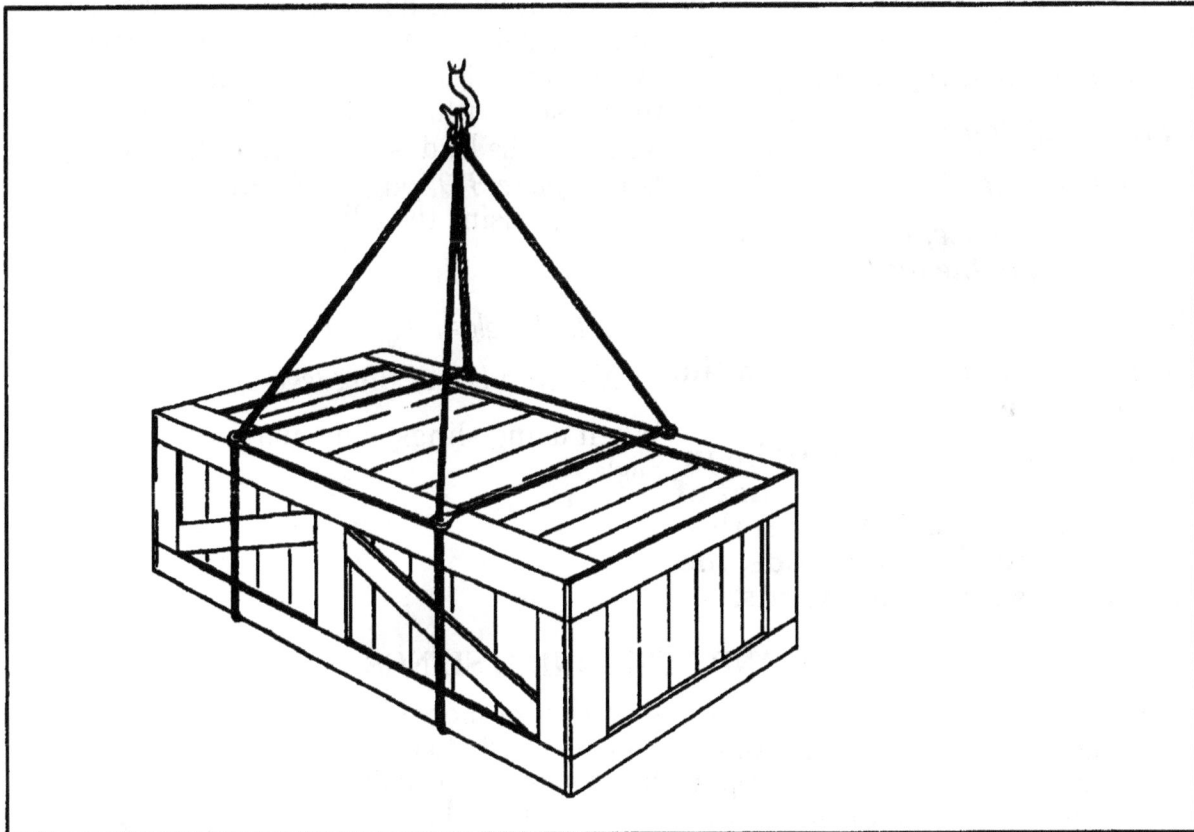

Figure 3-7. Use of spreaders in slings

STRESSES

Tables 3-3 through 3-5, pages 3-10 through 3-12, list the SWCs of ropes, chains, and wire-rope slings under various conditions. The angle of the legs of a sling must be considered as well as the strength of the material of which a sling is made. The lifting capacity of a sling is reduced as the angle of its legs to the horizontal is reduced (as the legs of a sling are spread) (see *Figure 3-7).* Thus, reducing the angle of the legs of a sling increases the tension on the sling legs. In determining the proper size of sling, you must determine the tension on each leg for each load (see *Figure 3-8, page 3-13).* You can compute this tension using the following formula:

$$T = \frac{W}{N} \times \frac{L}{V}$$

T = Tension in a single sling leg (which may be more than the weight of the load lifted)

W= Weight of the load to be lifted

N = Number of slings

L = Length of sling

V = Vertical distance, measured from the hook to the top of the load

NOTES:

1. L and V must be expressed in the same unit of measure.

2. The resulting tension will be in the same unit of measure as that of the weight of the load. Thus, if the weight of the load is in pounds, the tension will be given in pounds.

Example: Determine the tension of a single leg of a two-legged sling being used to lift a load weighing 1,800 pounds. The length of a sling is 8 feet and the vertical distance is 6 feet.

Solution:

$$T = \frac{W}{N} \times \frac{L}{V}$$

$$T = \frac{1,800}{2} \times \frac{8}{6} \qquad 1,200 \ pounds \ or \ tons$$

By knowing the amount of tension in a single leg, you can determine the appropriate size of fiber rope, wire rope, or chain. The SWC of a sling leg (keeping within the safety factors for slings) must be equal to or greater than the tension on a sling leg. If possible, keep the tension on each sling leg below that in the hoisting line to which the sling is attached. A particular angle formed by the sling legs with the horizontal where the tension within each sling leg equals the weight of the load is called the critical angle (see *Figure 3-9, page 3-13).* Approximate this angle using the following formula:

$$Critical \ angle = \frac{60}{N}$$

N = Number of sling legs

When using slings, stay above the critical angle.

INSPECTING AND CUSHIONING SLINGS

Inspect slings periodically and condemn them when they are no longer safe. Make the usual deterioration check for fiber ropes, wire ropes, chains, and hooks when you use them in slings. Besides the usual precautions, declare wire ropes used in slings unsafe if 4 percent or more of the wires are broken. Pad all objects to be lifted with wood blocks, heavy fabric, old rubber tires, or other cushioning material to protect the legs of slings from being damaged.

Table 3-3. SWCs for manila-rope slings (standard, three-strand manila-rope sling with a splice in each end)

Size		Single Sling	Double Sling			Quadruple Sling		
Circumference (inches)	Diameter (inches)	Vertical Lift (pounds)	60° Angle (pounds)	45° Angle (pounds)	30° Angle (pounds)	60° Angle (pounds)	45° Angle (pounds)	30° Angle (pounds)
3/4	1/4	108	187	153	108	374	306	216
1 1/8	3/8	241	418	341	241	836	683	482
1 1/2	1/2	475	822	672	475	1,645	1,345	950
2	5/8	791	1,370	1,119	791	2,740	2,238	1,585
2 1/4	3/4	970	1,680	1,375	970	3,360	2,750	1,940
2 3/4	7/8	1,382	2,395	1,945	1,382	4,790	3,890	2,764
3	1	1,620	2,805	2,290	1,620	5,610	4,580	3,240
3 1/2	1 1/8	2,160	3,740	3,060	2,160	7,480	6,120	4,320
3 3/4	1 1/4	2,430	4,205	3,437	2,430	8,410	6,875	4,860
4 1/2	1 1/2	3,330	5,770	4,715	3,330	11,540	9,430	6,660
5 1/2	1 3/4	4,770	8,250	6,750	4,770	16,500	13,500	9,540
6	2	5,580	9,670	7,900	5,580	19,340	15,800	11,160
7 1/2	2 1/2	8,366	14,500	11,850	8,366	29,000	23,700	16,732
9	3	11,520	19,950	16,300	11,520	39,900	32,600	23,040

Table 3-4. SWCs for chain slings (new wrought-iron chains)

Link Stock Diameter (Inches)	Single Sling	Double Sling			Quadruple Sling		
	Vertical Lift (pounds)	60° Angle (pounds)	45° Angle (pounds)	30° Angle (pounds)	60° Angle (pounds)	45° Angle (pounds)	30° Angle (pounds)
3/8	2,510	4,350	3,555	2,510	8,700	7,110	5,020
7/16	3,220	5,575	4,560	3,220	11,150	9,120	6,440
1/2	4,180	7,250	5,915	4,180	14,500	11,830	8,360
9/16	5,420	9,375	7,670	5,420	18,750	15,340	10,840
5/8	6,460	11,175	9,150	6,460	22,350	18,300	12,920
3/4	9,160	15,850	12,950	9,160	31,700	25,900	18,320
7/8	13,020	22,550	18,410	13,020	45,100	36,820	26,000
1	17,300	29,900	24,450	17,300	59,800	48,900	34,600
1 1/8	21,550	37,350	30,550	21,550	74,700	61,100	43,100
1 1/4	26,600	46,050	37,600	26,600	92,100	75,200	53,200
1 3/8	32,200	55,750	45,600	32,200	111,500	91,200	64,400
1 1/2	38,300	66,400	54,250	38,300	132,800	108,500	76,600
1 5/8	44,600	77,200	63,050	44,600	154,400	126,100	89,200
1 3/4	51,300	88,750	72,500	51,300	177,500	145,000	102,600
1 7/8	58,700	101,500	83,000	58,700	203,000	166,000	117,400
2	66,200	114,500	93,500	66,200	229,000	187,000	132,400

Table 3-5. SWCs for wire-rope slings (new, IPS wire rope)

Diameter (Inches)	Single Sling	Double Sling			Quadruple Sling		
	Vertical Lift (pounds)	60° Angle (pounds)	45° Angle (pounds)	30° Angle (pounds)	60° Angle (pounds)	45° Angle (pounds)	30° Angle (pounds)
1/4	1,096	1,899	1,552	1,096	3,798	3,105	2,192
5/16	1,690	2,925	2,390	1,690	5,850	4,780	3,380
3/8	2,460	4,260	3,485	2,460	8,520	6,970	4,920
7/16	3,560	6,170	5,040	3,560	12,340	10,080	7,120
1/2	4,320	7,475	6,105	4,320	14,950	12,210	8,640
9/16	5,460	9,450	7,725	5,460	18,900	15,450	10,920
5/8	6,650	11,500	9,400	6,650	23,000	18,800	13,300
3/4	9,480	16,400	13,400	9,480	32,800	26,800	18,960
7/8	12,900	22,350	18,250	12,900	44,700	36,500	25,800
1	16,800	29,100	23,750	16,800	58,200	47,500	33,600
1 1/8	21,200	36,700	30,000	21,200	73,400	60,000	42,400
1 1/4	26,000	45,000	36,800	26,000	90,000	73,600	52,000
1 3/8	32,000	55,400	45,250	32,000	110,800	90,500	64,000
1 1/2	37,000	64,000	52,340	37,000	128,000	104,700	74,000
1 5/8	41,800	72,400	59,200	41,800	144,800	118,400	83,600
1 3/4	49,800	86,250	70,500	49,800	172,500	141,000	99,600
2	62,300	107,600	88,050	62,300	215,200	176,100	124,600
2 1/4	82,900	143,500	117,400	82,900	287,000	234,800	165,800
2 1/2	101,800	176,250	144,000	101,800	352,500	288,000	203,600
2 3/4	122,500	212,000	173,500	122,500	424,000	347,000	245,000

Tension in a sling leg

$$T = \frac{W_L}{N} \times \frac{L}{V}$$

T = Tension in a single leg

W_L = Weight of load

N = Number of sling legs

L = Length of sling leg

V = Vertical distance of sling

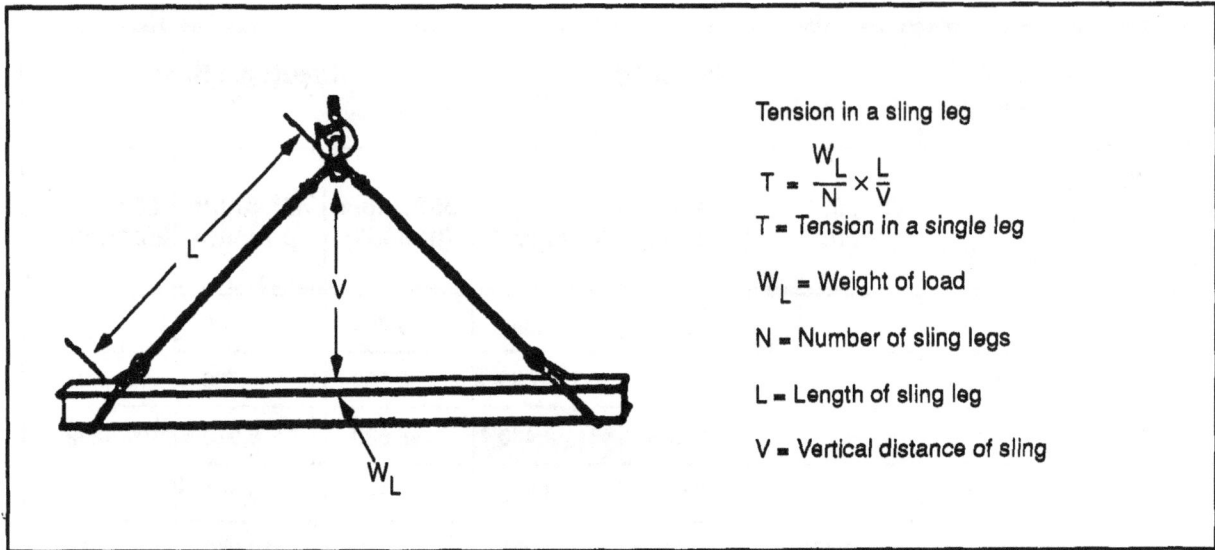

Figure 3-8. Computing tension in a sling

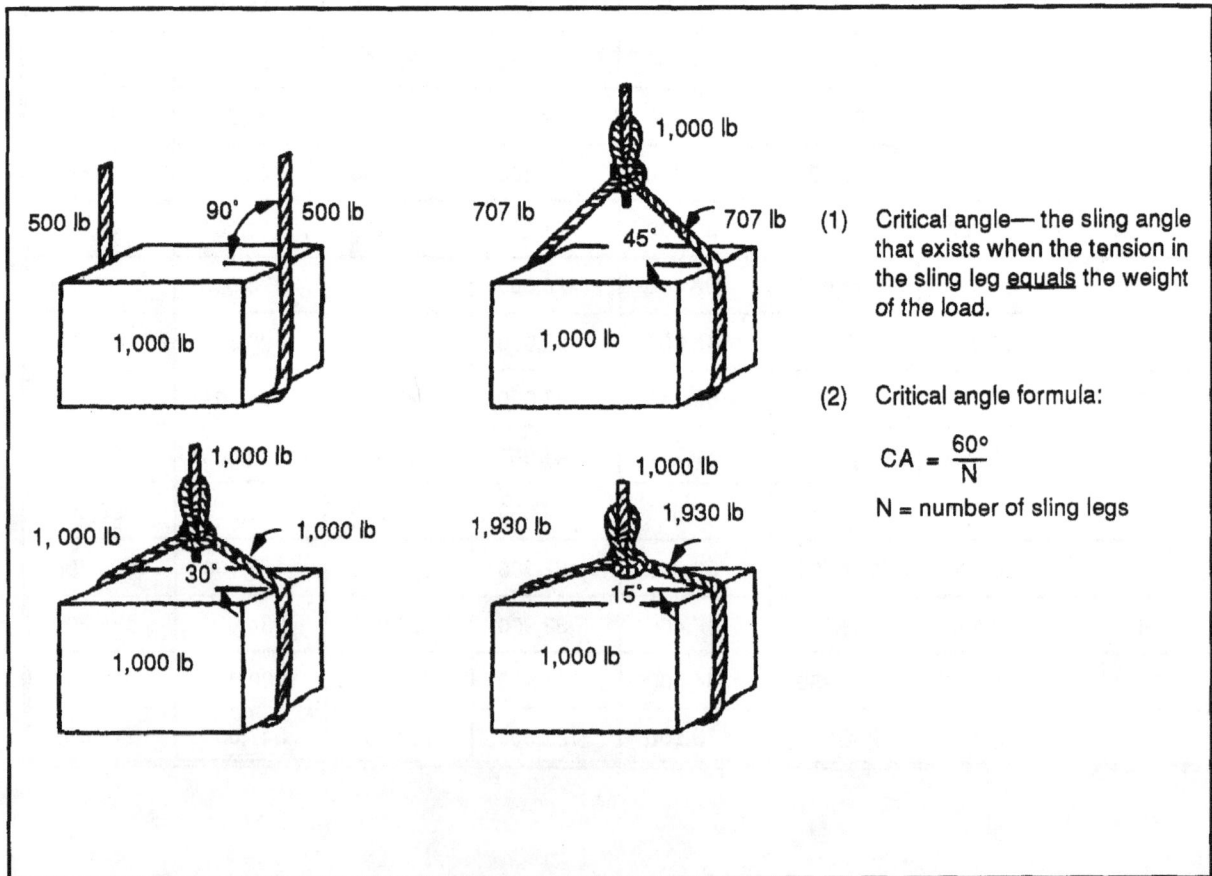

(1) Critical angle— the sling angle that exists when the tension in the sling leg <u>equals</u> the weight of the load.

(2) Critical angle formula:

$$CA = \frac{60°}{N}$$

N = number of sling legs

Figure 3-9. Sling angles

Section III. Blocks and Tackle Systems

A force is a push or pull. The push or pull that humans can exert depends on their weight and strength. To move any load heavier than the maximum amount a person can move, use a machine that multiplies the force exerted into a force capable of moving the load. The machine may be a lever, a screw, or a tackle system. The same principle applies to all of them. If you use a machine that exerts a force 10 times greater than the force applied to it, the machine has multiplied the force input by 10. The mechanical advantage (MA) of a machine is the amount by which the machine multiplies the force applied to it to lift or move a load. For example, if a downward push of 10 pounds on the left end of a lever will cause the right end of the lever to raise a load weighing 100 pounds, the lever is said to have a MA of 10.

A block consists of a wood or metal frame containing one or more rotating pulleys called sheaves (see Figure 3-10, A). A tackle is an assembly of ropes and blocks used to multiply forces (see Figure 3-10, B). The number of times the force is multiplied is the MA of the tackle. To make up a tackle system, lay out the blocks you are to use to be used and reeve (thread) the rope through the blocks. Every tackle system contains a fixed block attached to some solid support and may have a traveling block attached to the load. The single rope leaving the tackle system is called the fall line. The pulling force is applied to the fall line, which may be led through a leading block. This is an additional block used to change the direction of pull.

BLOCKS

Blocks are used to reverse the direction of the rope in the tackle. Blocks take their names from—

- The purpose for which they are used.
- The places they occupy.
- A particular shape or type of construction (see Figure 3-11).

TYPES OF BLOCKS

Blocks are designated as single, double, or triple, depending on the number of sheaves.

Snatch Block

This is a single sheave block made so that the shell opens on one side at the base of the hook to permit a rope to be slipped over the sheave without threading the end of it through the block. Snatch blocks ordinarily

are used where it is necessary to change the direction of the pull on the line.

Traveling Block

A traveling block is attached to the load that is being lifted and moves as the load is lifted.

Standing Block

This block is fixed to a stationary object.

Leading Blocks

Blocks used in the tackle to change the direction of the pull without affecting the MA of the system are called leading blocks (see Figure 3-12, page 3-16). In some tackle systems, the fall line leads off the last block in a direction that makes it difficult to apply the motive force required. A leading block is used to correct this. Ordinarily, a

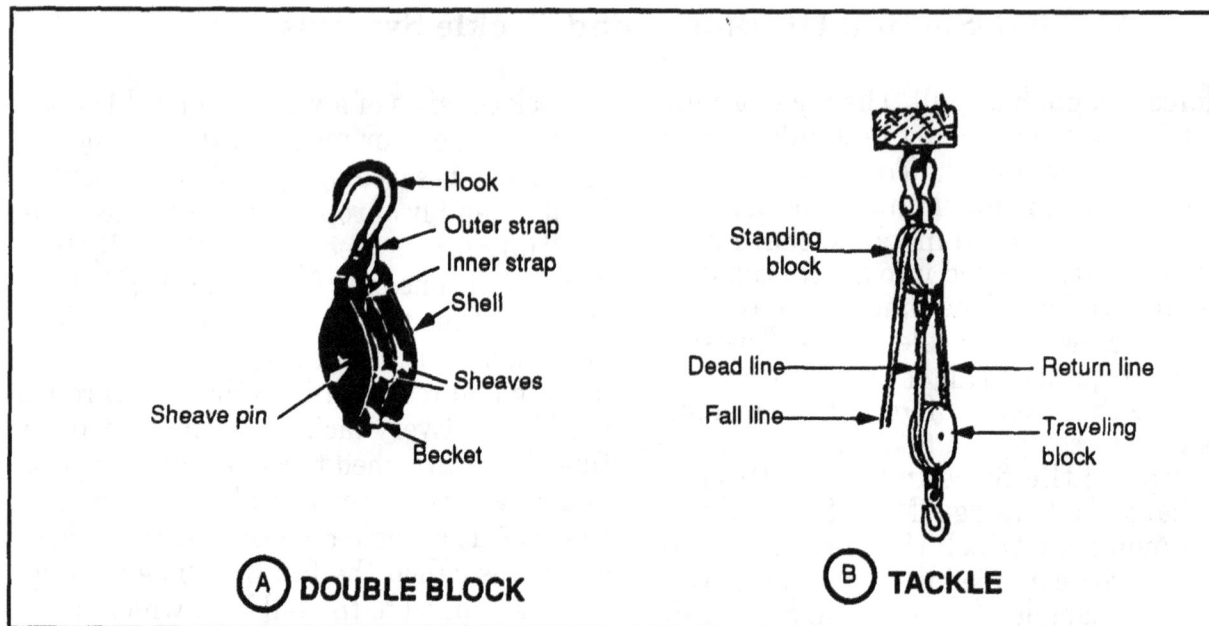

Figure 3-10. Double block and tackle system

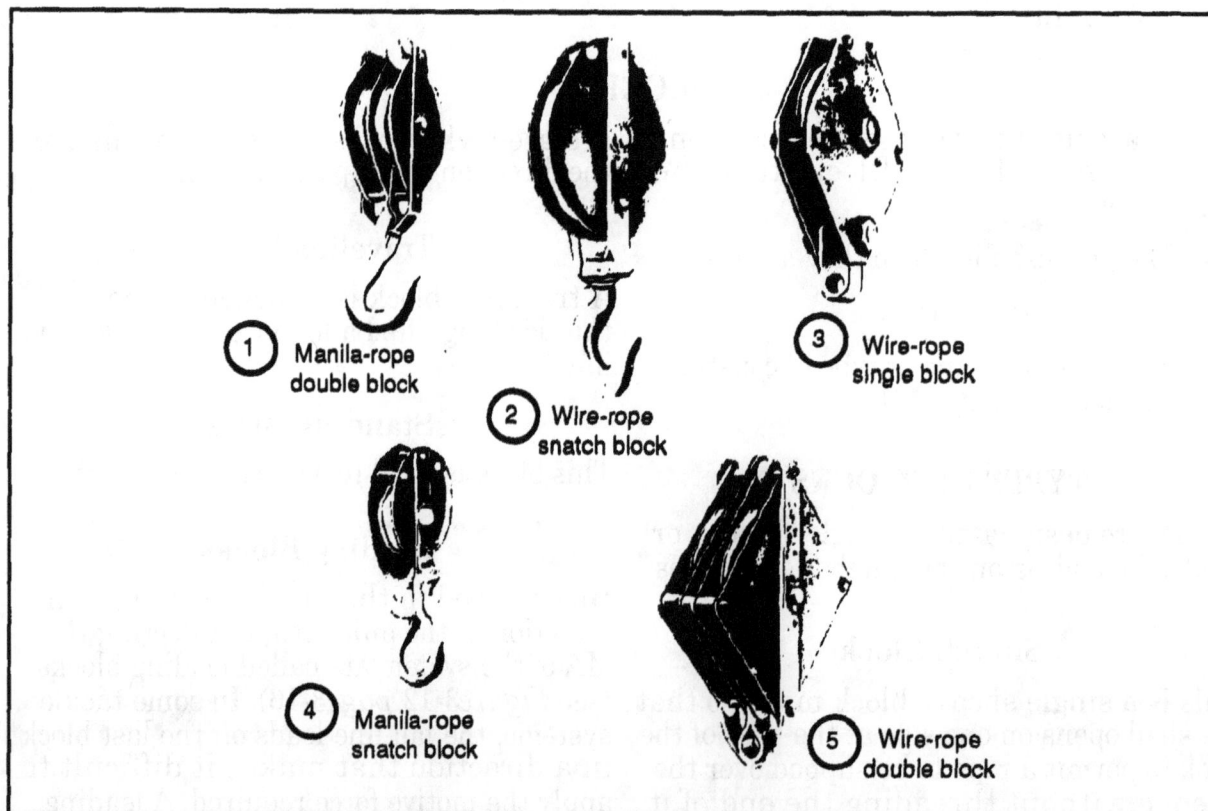

Figure 3-11. Types of blocks

snatch block is used as the leading block. This block can be placed at any convenient position. The fall line from the tackle system is led through the leading block to the line of most direct action.

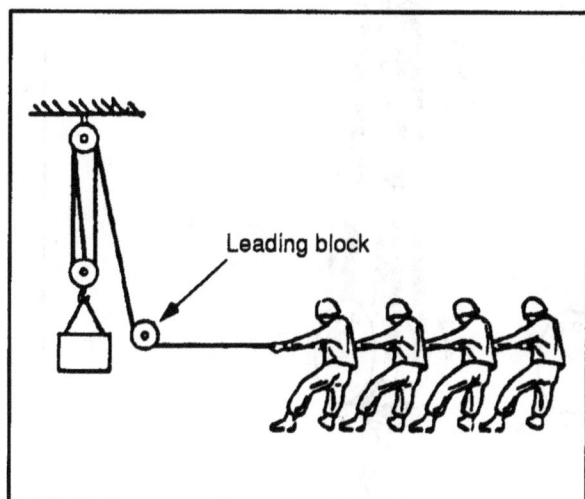

Figure 3-12. Use of leading block

REEVING BLOCKS

To prepare blocks for use, reeve, or pass a rope through, it. To do this, lay out the blocks on a clean and level surface other than the ground to avoid getting dirt into the operating parts, *Figure 3-13* shows the reeving of single and double blocks. In reeving triple blocks, it is imperative that you put the hoisting strain at the center of the blocks to prevent them from being inclined under the strain (see *Figure 3-14*) If the blocks do incline, the rope will drag across the edges of the sheaves and the shell of the block and cut the fibers. Place the blocks so that the sheaves in one block are at right angles to the sheaves in the other block. You may lay the coil of rope beside either block. Pass the running end over the center sheave of one block and back to the bottom sheave of the other block. Then pass it over one of the side sheaves of the first block. In selecting which side sheave to pass the rope

over, remember that the rope should not cross the rope leading away from the center sheave of the first block. Lead the rope over the top sheave of the second block and back to the remaining side sheave of the first block. From this point, lead the rope to the center sheave of the second block and back to the becket of the first block. Reeve the rope through the blocks so that no part of the rope chafes another part of the rope.

Twisting of Blocks

Reeve blocks so as to prevent twisting. After reeving the blocks, pull the rope back and forth through the blocks several times to allow the rope to adjust to the blocks. This reduces the tendency of the tackle to twist under a load. When the ropes in a tackle system become twisted, there is an increase in friction and chafing of the ropes, as well as a possibility of jamming the blocks. When the hook of the standing block is fastened to the supporting member, turn the hook so that the fall line leads directly to the leading block or to the source of motive power. It is very difficult to prevent twisting of a traveling block. It is particularly important when the tackle is being used for a long pull along the ground, such as in dragging logs or timbers.

Antiwisting Devices

One of the simplest antitwisting devices for such a tackle is a short iron rod or a piece of pipe lashed to the traveling block (see *Figure 3-15, page 3-18*) You can lash the antitwisting rod or pipe to the shell of the block with two or three turns of rope. If it is lashed to the becket of the block, you should pass the rod or pipe between the ropes without chafing them as the tackle is hauled in.

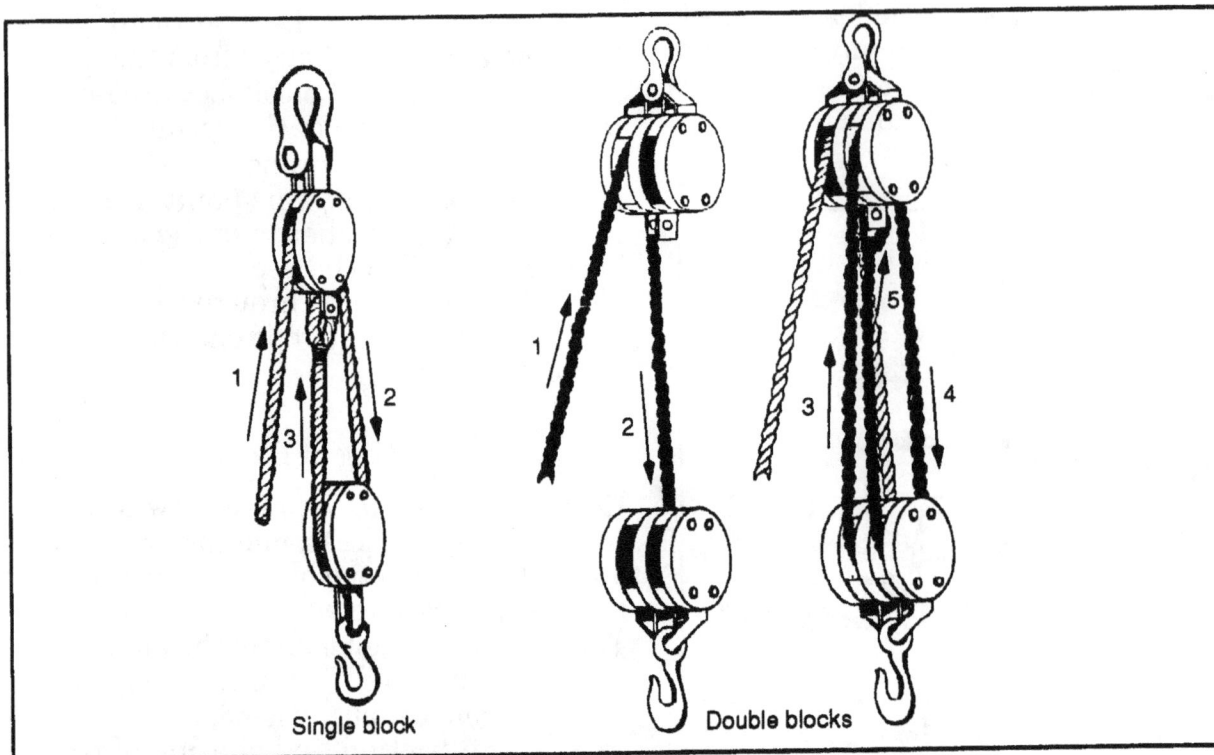

Figure 3-13. Reeving single and double blocks

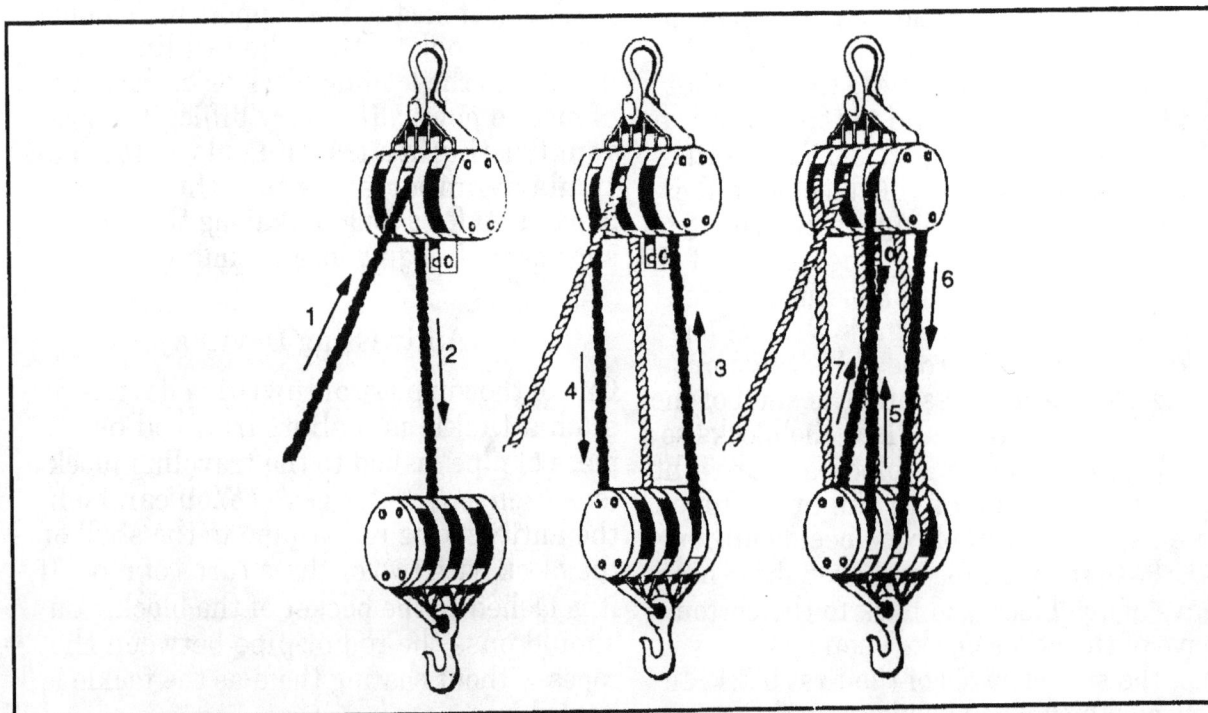

Figure 3-14. Reeving triple blocks

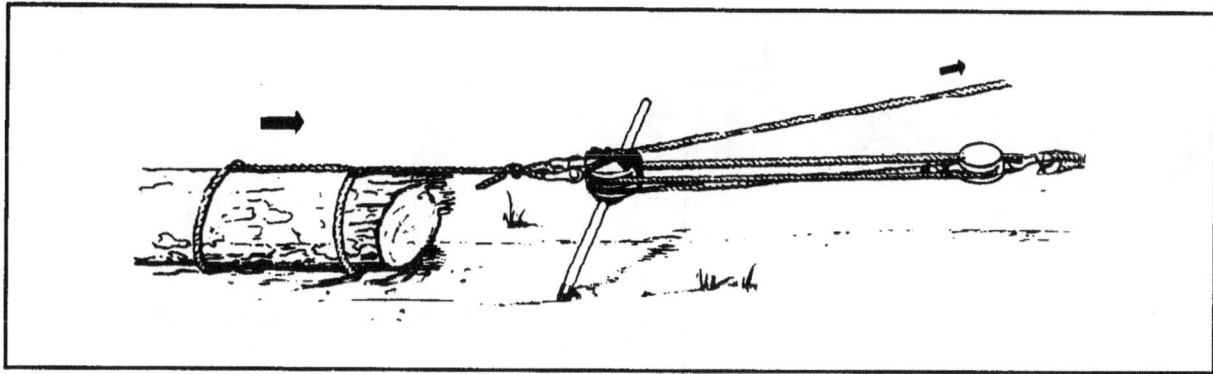

Figure 3-15. Antitwisting rod or pipe

TACKLE SYSTEMS

Tackle systems may be either simple or compound.

SIMPLE TACKLE SYSTEMS

A simple tackle system uses one rope and one or more blocks. To determine the MA of a simple system, count the number of lines supporting the load (or the traveling block) (see *Figure 3-16).* In counting, include the fall line if it leads out of a traveling block. In a simple tackle system, the MA always will be the same as the number of lines supporting the load. As an alternate method, you can determine the MA by tracing the forces through the system. Begin with a unit force applied to the fall line. Assume that the tension in a single rope is the same throughout and therefore the same force will exist in each line. Total all the forces acting on the load or traveling block. The ratio of the resulting total force acting on the load or traveling block to the original unit force exerted on the fall line is the theoretical MA of the simple system.

Figure 3-17 shows examples of two methods of determining the ratio of a simple tackle system. They are—

- Method I-counting supporting lines.
- Method II—unit force.

Method I—Counting Supporting Lines

There are three lines supporting the traveling block, so the theoretical MA is 3:1.

Method II—Unit Force

Assuming that the tension on a single rope is the same throughout its length, a unit force of 1 on the fall line results in a total of 3 unit forces acting on the traveling block. The ratio of the resulting force of 8 on the traveling block to the unit force of 1 on the fall line gives a theoretical MA of 3:1.

COMPOUND TACKLE SYSTEMS

A compound tackle system uses more than one rope with two or more blocks (see *Figure 3-18, page* 3-20). Compound systems are made up of two or more simple systems. The fall line from one simple system is fastened to a hook on the traveling block of another simple system, which may include one or more blocks. In compound systems, you can best determine the MA by using the unit-force method. Begin by applying a unit force to the fall line. Assume that the tension in a single rope is the same throughout and therefore the same force will exist in each line. Total all the forces acting on the

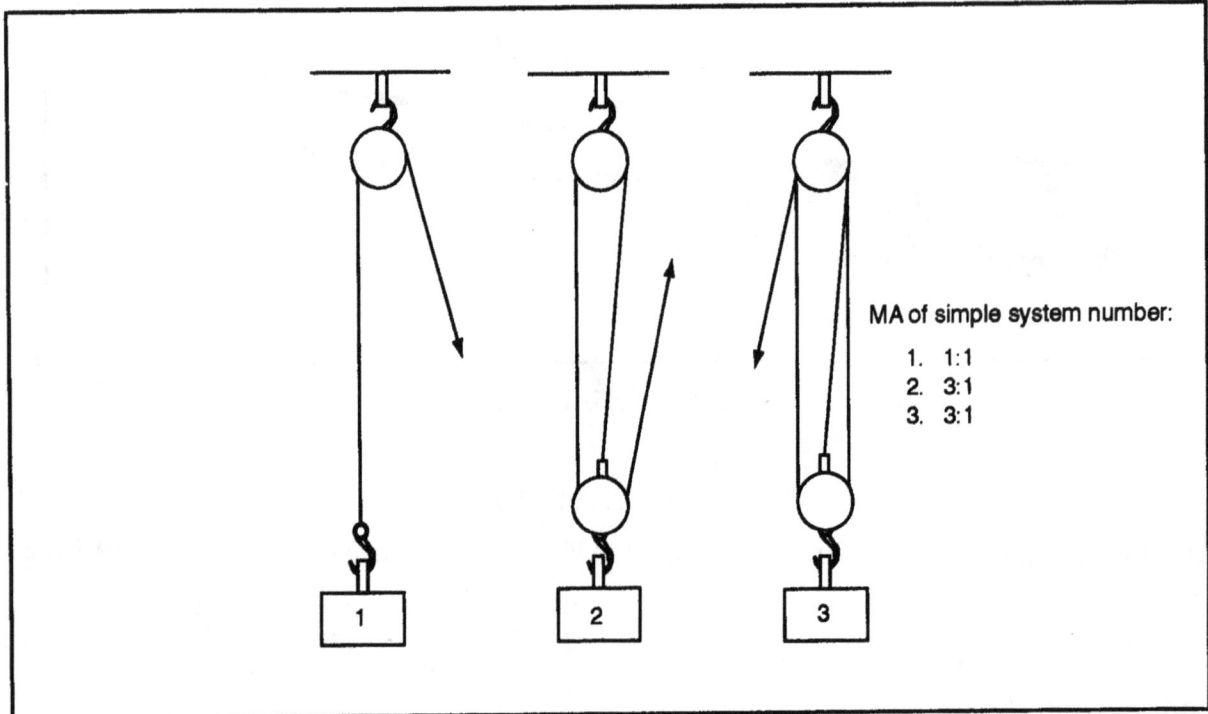

MA of simple system number:
1. 1:1
2. 3:1
3. 3:1

Figure 3-16. Simple tackle systems

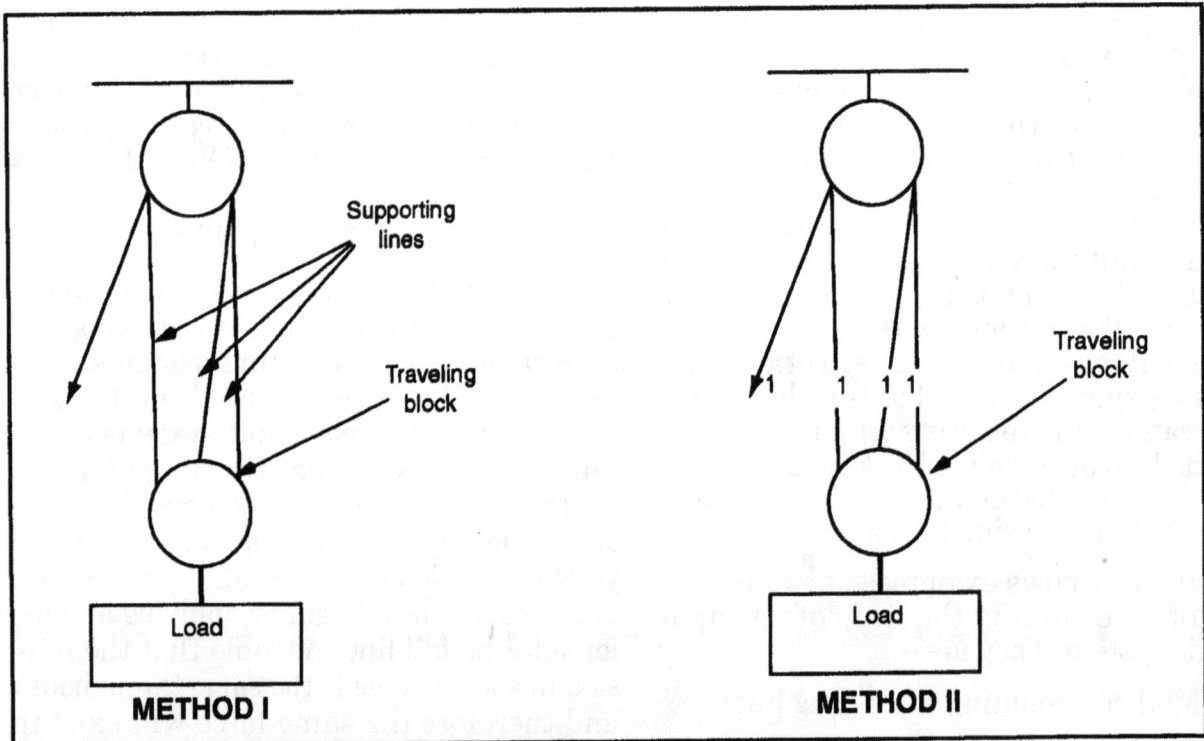

Figure 3-17. Determining ratio of a simple tackle

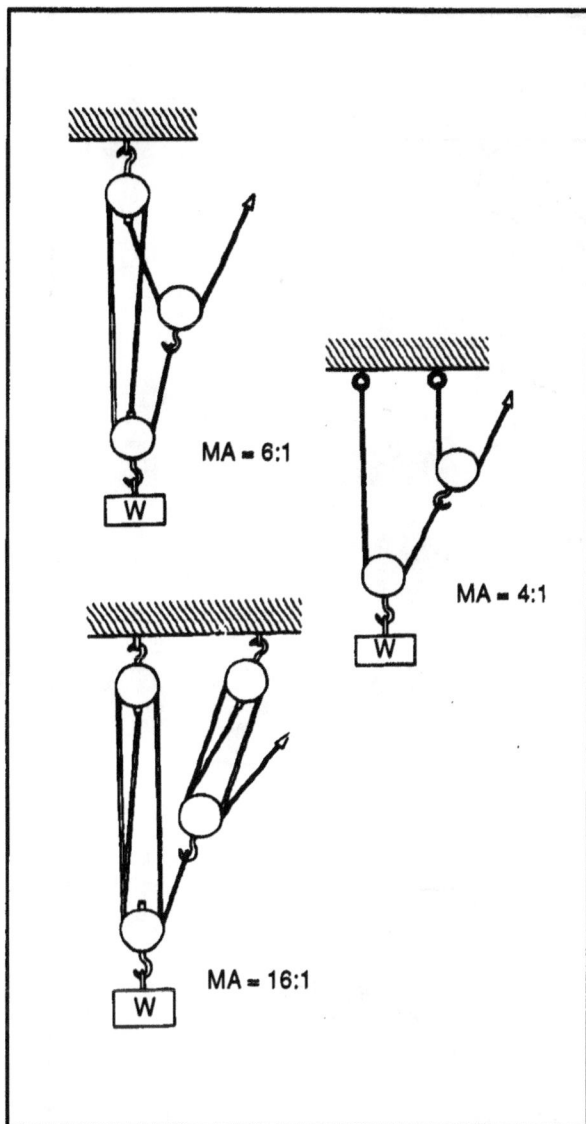

Figure 3-18. Compound tackle systems

two methods of determining the ratio of a compound tackle system. They are—

- Method I—unit force.
- Method II—multiplying mechanical advantages of simple systems.

Method I—Unit Force

As in method II of simple tackle systems, a unit force of 1 on the fall line results in 4 unit forces acting on the traveling block of tackle system A. Transferring the unit force of 4 into the fall line of simple system B results in a total of 16 unit forces (4 lines with 4 units of force in each) acting on the traveling block of tackle system B. The ratio of 16 unit forces on the traveling block carrying the load to a 1 unit force on the fall line gives a theoretical MA of 16:1.

Method II—Multiplying MAs of Simple Systems

The number of lines supporting the traveling blocks in systems A and B is equal to 4. The MA of each simple system is therefore equal to 4:1. You can then determine the MA of the compound system by multiplying together the MA of each simple system for a resulting MA of 16:1.

FRICTION

There is a loss in any tackle system because of the friction created by—

- The sheave rolling on the pin, the ropes rubbing together.
- The rope rubbing against the sheave.

This friction reduces the total lifting power; therefore, the force exerted on the fall line must be increased by some amount to overcome the friction of the system to lift the load. Each sheave in the tackle system can be expected to create a resistance equal to about 10 percent of the weight of the load.

traveling block and transfer this force into the next simple system. The ratio of the resulting total force acting on the load or traveling block to the original unit force exerted on the fall line is the theoretical MA of the compound system. Another method, which is simpler but less accurate in some cases, is to determine the MA of each simple system in the compound system and multiplying these together to obtain the total MA. *Figure 3-19* shows examples of the

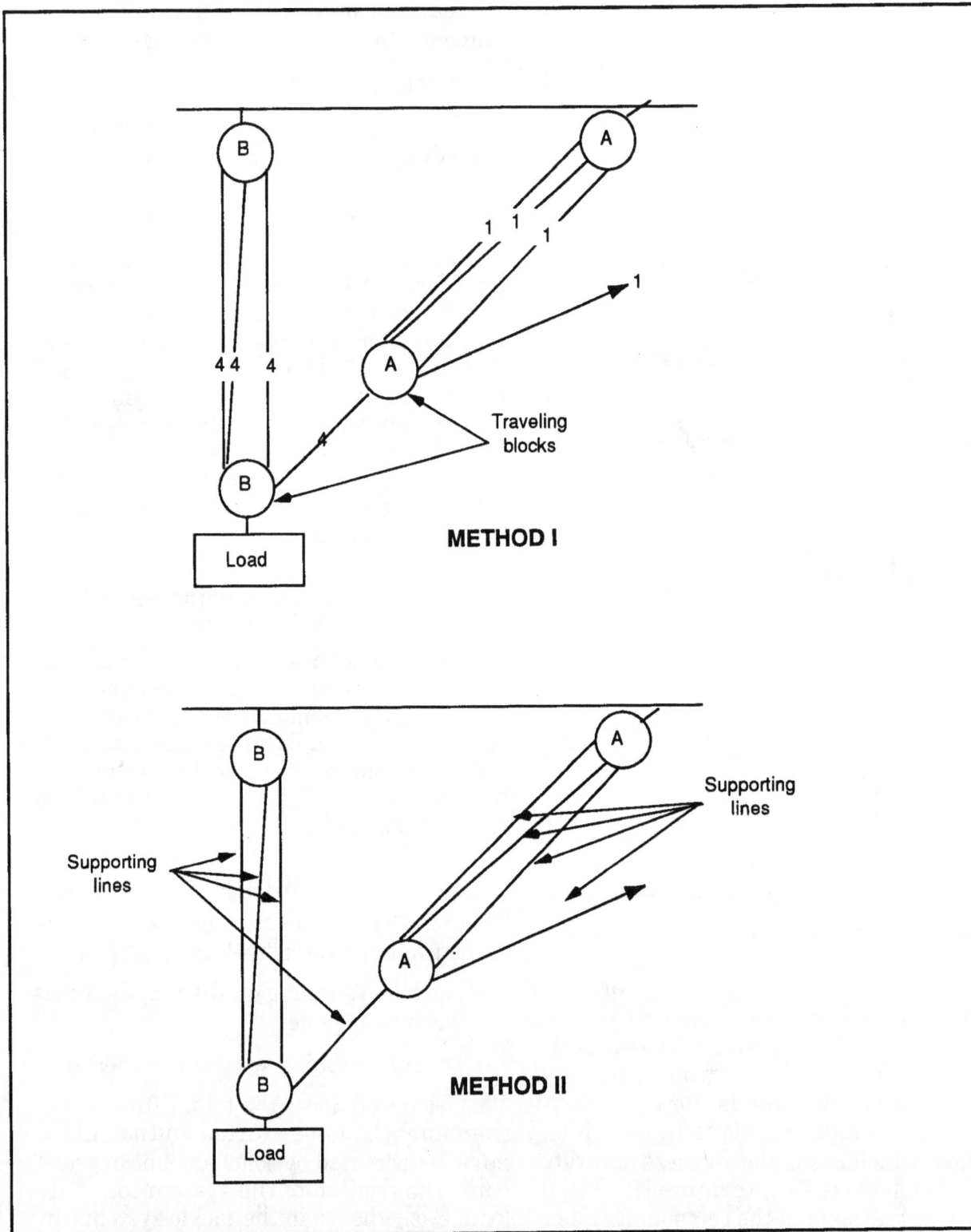

Figure 3-19. Determining ratio of a compound tackle system

Example—A load weighing 5,000 pounds is lifted by a tackle system that has a MA of 4:1. The rope travels over four sheaves that produce a resistance of 40 percent of 5,000 pounds or 2,000 pounds (5,000 x 0.40). The actual pull that would be required on the fall line of the tackle system is equal to the sum of the weight of the load and the friction in the tackle system divided by the theoretical.

MA of the tackle system. The actual pull required on the fall line would be equal to the sum of 5,000 pounds (load) and 2,000 pounds (friction) divided by 4 (MA) or 1,750 pounds.

There are other types of resistance that may have to be considered in addition to tackle resistance. *FM* 20-22 presents a thorough discussion of resistance.

Section IV. Chain Hoists and Winches

In all cases where manpower is used for hoisting, the system must be arranged to consider the most satisfactory method of using that source of power. More men can pull on a single horizontal line along the ground than on a single vertical line. On a vertical pull, men of average weight can pull about 100 pounds per man and about 60 pounds per man on a horizontal. If the force required on the fall line is 300 pounds or less, the fall line can lead directly down from the upper block of a

tackle vertical line. If 300 pounds times the MA of the system is not enough to lift a given load, the tackle must be rigged again to increase the MA, or the fall line must be led through a leading block to provide a horizontal pull. This will permit more people to pull on the line. Similarly, if a heavy load is to be lifted and the fall line is led through a leading block to a winch mounted on a vehicle, the full power available at the winch is multiplied by the MA of the system.

CHAIN HOISTS

Chain hoists provide a convenient and efficient method for hoisting by hand under particular circumstances (see *Figure 3-20*). The chief advantages of chain hoists are that—

- The load can remain stationary without requiring attention.
- One person can operate the hoist to raise loads weighing several tons.

The slow lifting travel of a chain hoist permits small movements, accurate adjustments of height, and gentle handling of loads. A retched-handle pull hoist is used for short horizontal pulls on heavy objects (see *Figure 3-21*). Chain hoists differ widely in their MA, depending on their rated capacity which may vary from 5 to 250.

TYPES OF CHAIN HOISTS

The three general types of chain hoists for

vertical operation are the spur gear, screw gear, and differential.

Spur-Gear Chain Hoist

This is the most satisfactory chain hoist for ordinary operation where a minimum number of people are available to operate the hoist and the hoist is to be used frequently. This type of chain hoist is about 85 percent efficient.

Screw-Gear Chain Hoist

The screw-gear chain hoist is about 50 percent efficient and is satisfactory where less frequent use of the chain hoist is involved.

Differential Chain Hoist

The differential chain hoist is only about 35 percent efficient but is satisfactory for occasional use and light loads.

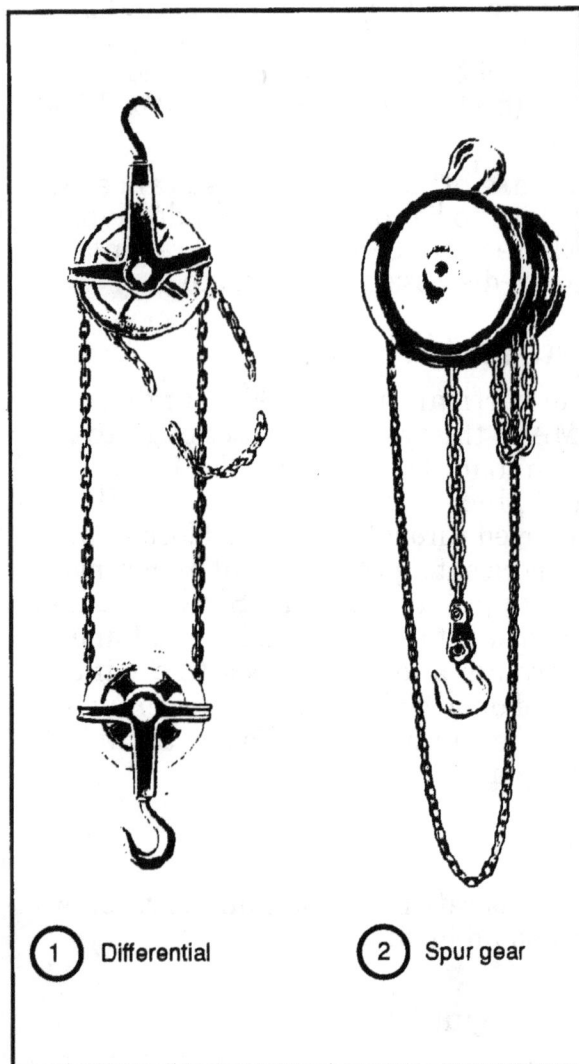

Figure 3-20. Chain hoists

LOAD CAPACITY

Chain hoists are usually stamped with their load capacities on the shell-of the upper block. The rated load capacity will run from one-half of a ton upward. Ordinarily, chain hoists are constructed with their lower hook as the weakest part of the assembly. This is done as a precaution so that the lower hook will be overloaded before the chain hoist is overloaded. The lower hook will start to spread under overload, indicating to the operator that he is approaching the overload point of the chain hoist. Under ordinary

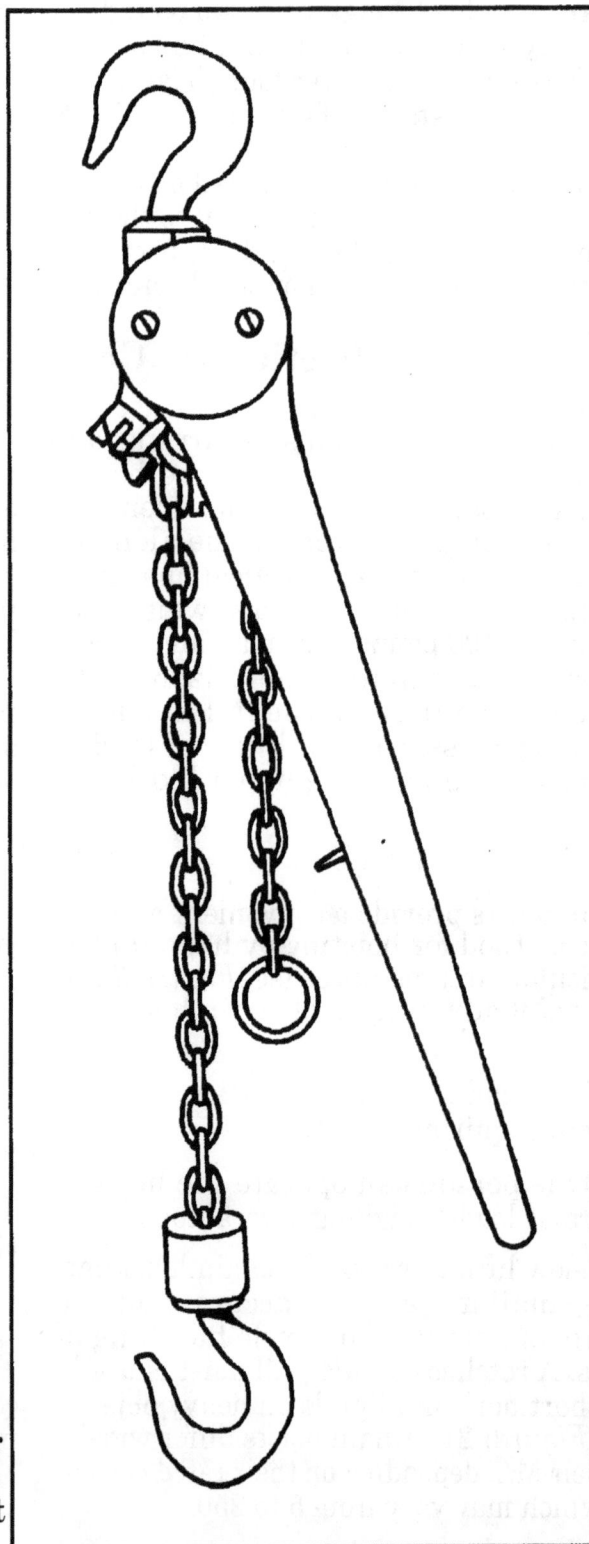

Figure 3-21. Ratched-handle chain hoist

circumstances, the pull exerted on a chain hoist by one or two men will not overload the hoist. Inspect chain hoists at frequent intervals. Any evidence of spreading of the hook or excessive wear is sufficient cause to replace the hook. If the links of the chain are distorted, it indicates that the chain hoist has been heavily over-loaded and is probably unsafe for further use. Under such circumstances, the chain hoist should be condemned.

WINCHES

Vehicular-mounted and engine-driven winches are used with tackles for hoisting (see *Figure 3-22)*. There are two points to consider when placing a power-driven winch to operate hoisting equipment. They are—

- The angle with the ground that the hoisting line makes at the drum of the hoist.
- The fleet angle of the hoisting line winding on the drum (see *Figure*3-23).

The distance from the drum to the first sheave of the system is the controlling factor in the fleet angle. When using vehicular-mounted winches, place the vehicle in a position that lets the operator watch the load being hoisted. A winch is most effective when the pull is exerted on the bare drum of the winch. When a winch is rated at a capacity, that rating applies only as the first layer of cable is wound onto the drum. The winch capacity is reduced as each layer of cable is wound onto the drum because of the change in leverage resulting from the increased diameter of the drum. The capacity of the winch may be reduced by as much as 50 percent when the last layer is being wound onto the drum.

GROUND ANGLE

If the hoisting line leaves the drum at an angle upward from the ground, the result-ing pull on the winch will tend to lift it clear of the ground. In this case, a leading block must be placed in the system at some distance from the drum to change the direc-tion of the hoisting line to a horizontal or downward pull. The hoisting line should be overwound or underwound on the drum as may be necessary to avoid a reverse bend.

FLEET ANGLE

The drum of the winch is placed so that a line from the last block passing through the center of the drum is at right angles to the axis of the drum. The angle between this line and the hoisting line as it winds on the drum is called the fleet angle (see*Figure 3-23)*. As the hoisting line is wound in on the drum, it moves from one flange to the other so that the fleet angle changes during the hoisting process. The fleet angle should not be permitted to exceed 2 degrees and should be kept below this, if possible. A 1 1/ 2-degree maximum angle is satisfactory and will be obtained if the distance from the drum to the first sheave is 40 inches for each inch from the center of the drum to the flange. The wider the drum of the hoist the greater the lead distance must be in placing the winch.

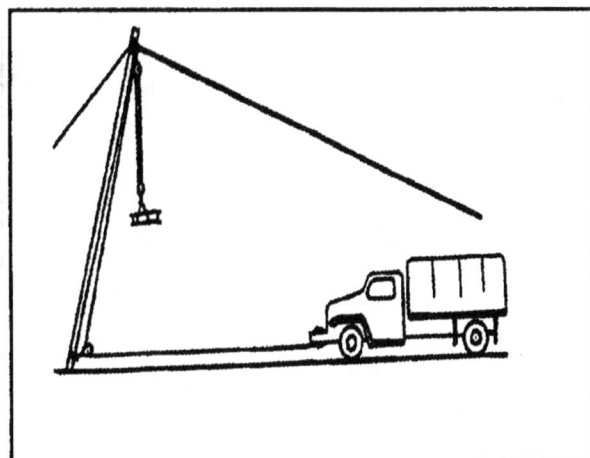

Figure 3-22. Using a vehicular winch for hoisting

SPANISH WINDLASS

In the absence of mechanical power or an appropriate tackle, you may have to use makeshift equipment for hoisting or pulling. You can use a Spanish windlass to move a load along the ground, or you can direct the horizontal pull from the windlass through the blocks to provide a vertical pull on a load. In making a Spanish windlass, fasten a rope between the load you are to move and an anchorage some distance away. Place a short spar vertically beside this rope, about halfway between the anchorage and the load (see *Figure* 3-24, *page* 3-26). This spar may be a pipe or a pole, but in either case it should have as large a diameter as possible. Make a loop in the rope and wrap it partly around the spar. Insert the end of a horizontal rod through this loop. The horizontal rod should be a stout pipe or bar long enough to provide leverage. It is used as a lever to turn the vertical spar. As the vertical spar turns, the rope is wound around it, which shortens the line and pulls on the load. Make sure that the rope leaving the vertical spar is close to the same level on both sides to prevent the spar from tipping over.

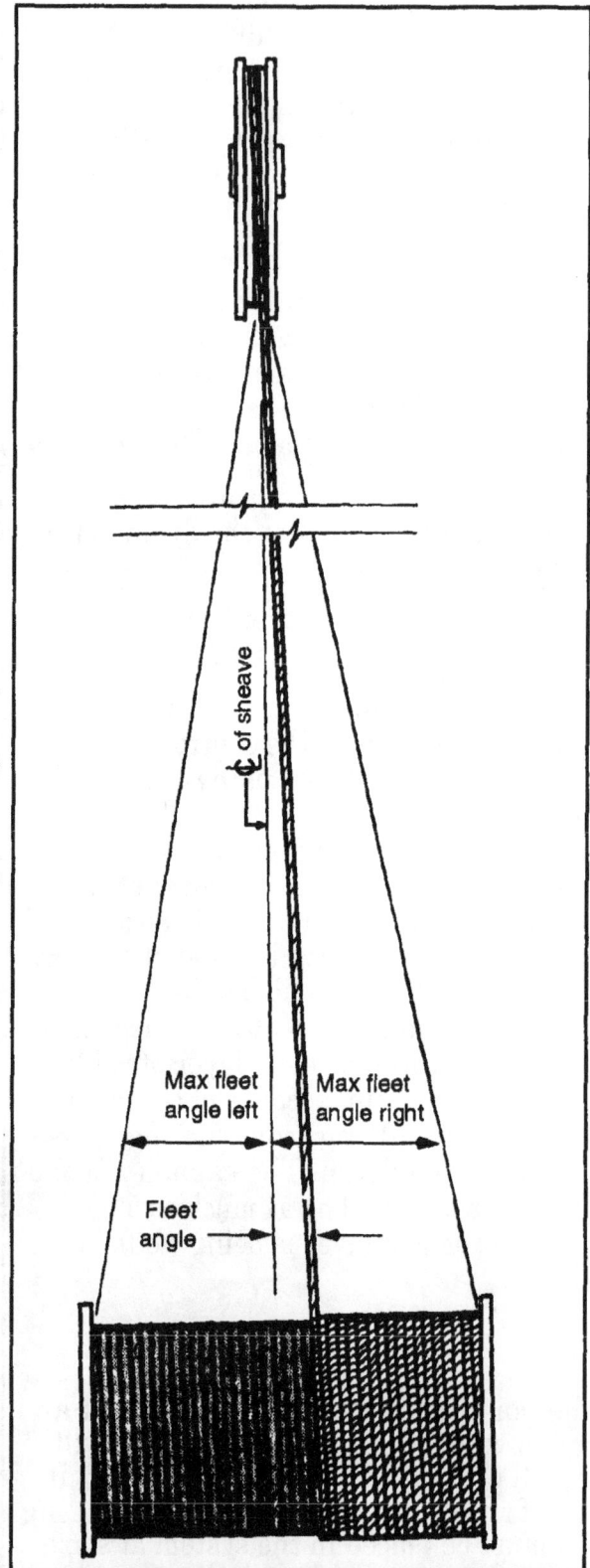

Figure 3-23. Fleet angle

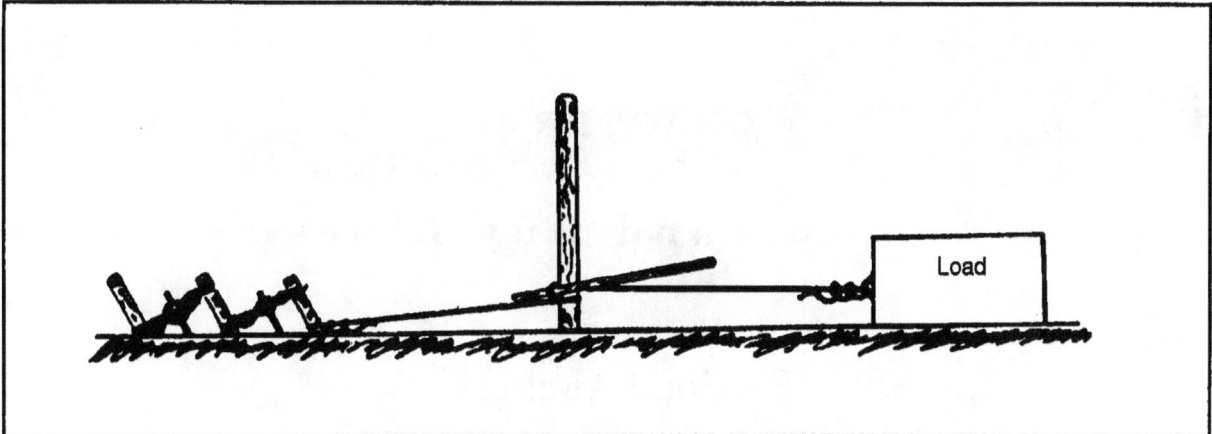

Figure 3-24. Spanish windlass

CHAPTER 4

Anchors and Guy Lines

Section I. Anchors

When heavy loads are handled with a tackle, it is necessary to have some means of anchorage. Many expedient rigging installations are supported by combining guy lines and some type of anchorage system. Anchorage systems may be either natural or man-made. The type of anchorage to be used depends on the time and material available and on the holding power required. Whenever possible, natural anchorages should be used so that time, effort, and material can be conserved. The ideal anchorage system must be of sufficient strength to support the breaking strength of the attached line. Lines should always be fastened to anchorages at a point as near to the ground as possible. The principal factor in the strength of most anchorage systems is the area bearing against the ground.

NATURAL ANCHORS

Trees, stumps, or rocks can serve as natural anchorages for rapid work in the field. Always attach lines near the ground level on trees or stumps (see *Figure 4-1*. Avoid dead or rotten trees or stumps as an anchorage because they are likely to snap suddenly when a strain is placed on the line. It is always advisable to lash the first tree or stump to a second one to provide added support. Place a transom between two trees to provide a stronger anchorage than a single tree (see *Figure 4-2, page 4-2*. When using rocks as natural anchorages, examine the rocks carefully to be sure that they are large enough and firmly embedded in the ground (see *Figure 4-3, page 4-2*. An outcropping of rock or a heavy boulder buried partially in the ground will serve as a satisfactory anchor.

Figure 4-1. Natural anchorage (tree)

Figure 4-2. Natural anchorage (trees and transom)

Figure 4-3. Natural anchorage (rock)

MAN-MADE ANCHORS

You must construct man-made anchors when natural anchors are not available. These include—

- Rock anchors.
- Picket holdfasts.
- Combination holdfasts.
- Deadmen.

ROCK ANCHORS

Rock anchors have an eye on one end and a threaded nut, an expanding wedge, and a stop nut on the other end (see *Figure 4-4*). To construct a rock anchor, insert the threaded end of the rock anchor in the hole with the nut's relation to the wedge as shown in *Figure 4-4* After placing the anchor, insert a crowbar through the eye of the rock anchor and twist it. This causes the threads to draw the nut up against the wedge and force the wedge out against the sides of the hole in the rock. The wedging action is strongest under a direct pull; therefore, always set rock anchors so that the pull is in a direct line with the shaft of the anchor. Drill the holes for rock anchors 5 inches deep. Use a 1-inch-diameter drill for hard rock and a 3/4-inch-diameter drill for soft rock. Drill the hole as neatly as possible so that the rock anchor can develop the maximum strength. In case of extremely soft rock, it is better to use some other type of anchor because the wedging action may not provide sufficient holding power.

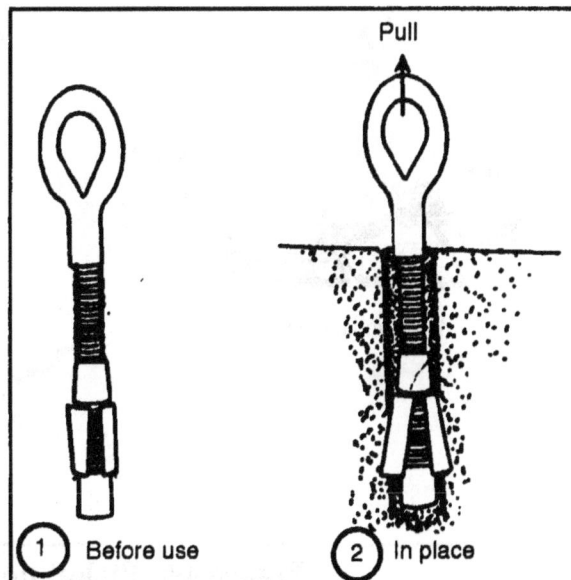

Figure 4-4. Rock anchor

PICKET HOLDFASTS

A single picket, either steel or wood, can be driven into the ground as an anchor. The holding power depends on the—

- Diameter and kind of material used.
- Type of soil.
- Depth and angle in which the picket is driven.
- Angle of the guy line in relation to the ground.

Table 4-1 lists the holding capacities of the various types of wooden picket holdfasts. *Figure 4-5* shows the various picket holdfasts.

Table 4-1. Holding power of picket holdfast in loamy soil

Holdfast	Pounds
Single picket	700
1-1 picket holdfast	1,400
1-1-1 picket holdfast	1,800
2-1 picket holdfast	2,000
3-2-1 picket holdfast	4,000

Note: Wet earth factors:
Clay and gravel mixtures - 0.9
Riven clay and sand - 0.5

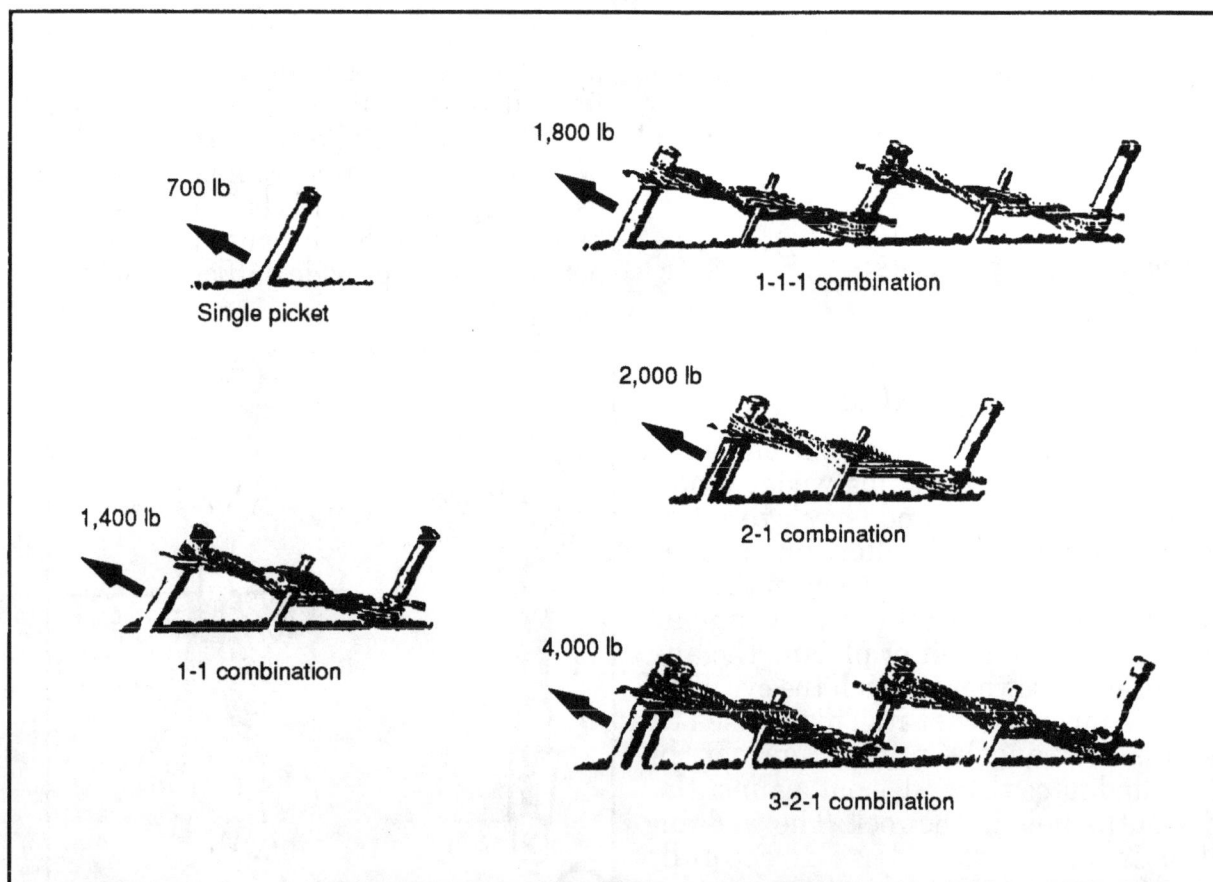

Figure 4-5. Picket holdfasts (loamy soil)

Single Wooden Pickets

Wooden stakes used for pickets should be at least 3 inches in diameter and 5 feet long. Drive the picket 3 feet into the ground at an angle of 15 degrees from the vertical and inclined away from the direction of pull (see *Figure 4-6*).

Multiple Wooden Pickets

You can increase the strength of a holdfast by increasing the area of the picket bearing against the ground. Two or more pickets driven into the ground, spaced 3 to 6 feet apart and lashed together to distribute the load, are much stronger than a single picket (see *Figure 4-6, A)*. To construct the lashing, tie a clove hitch to the top of the first picket with four to six turns around the first and second pickets, leading from the top of the first picket to the bottom of the second picket (see *Figure 4-6, B)*. Then fasten the rope to the second picket with a clove hitch just above the turns. Put a stake between the rope turns to tighten the rope by twisting the stake and then driving it into the ground (see *Figure 4-6, C)*. This distributes the load between the pickets. If you use more than two pickets, make a similar lashing between the second and third pickets (see *Figure 4-6, D)*. If you use wire rope for lashing, make only two complete turns around each pair of pickets. If neither fiber rope nor wire rope is available for lashing, place boards from the top of the front picket to the bottom of the second picket and nail them onto each picket (see *Figure 4-7)*. As you place pickets farther away from the front picket, the load to the rear pickets is distributed more unevenly. Thus, the prin-

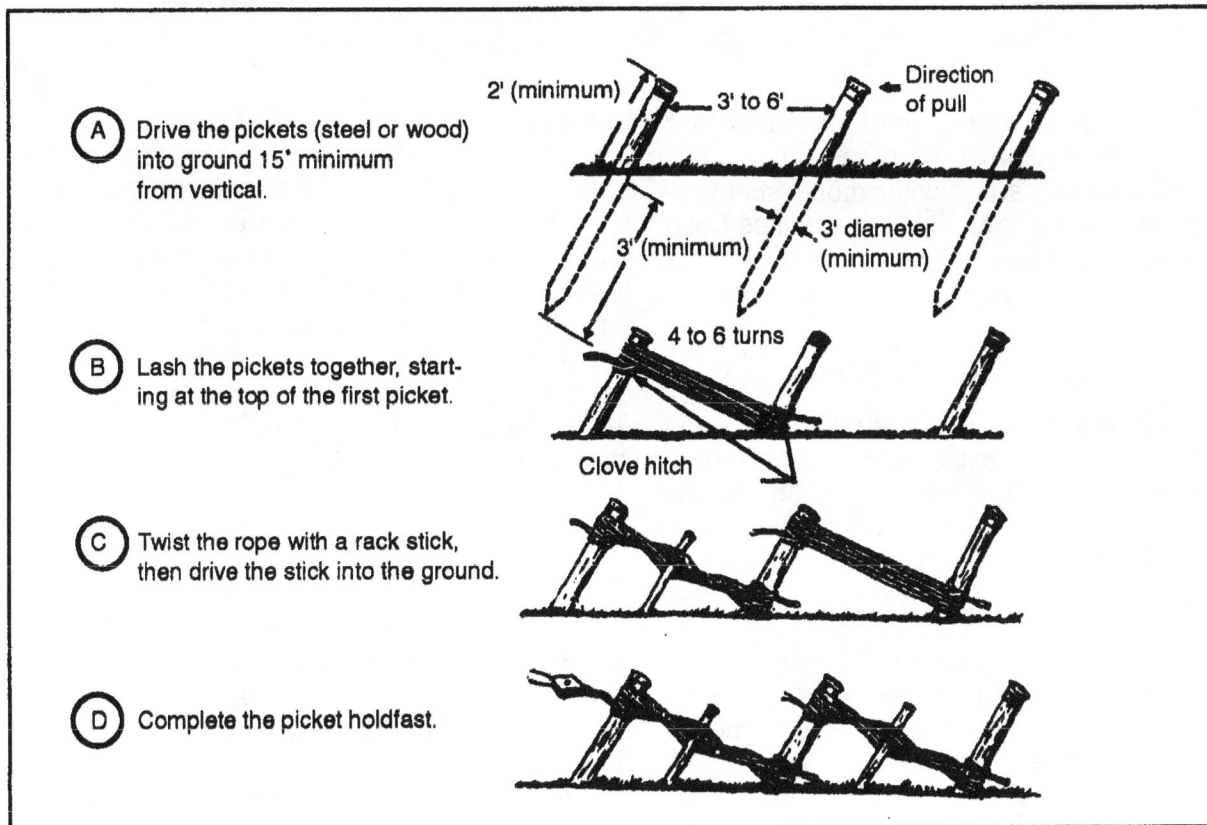

A Drive the pickets (steel or wood) into ground 15° minimum from vertical.

2' (minimum)   3' to 6'   Direction of pull

3' (minimum)   3' diameter (minimum)

B Lash the pickets together, starting at the top of the first picket.

4 to 6 turns

Clove hitch

C Twist the rope with a rack stick, then drive the stick into the ground.

D Complete the picket holdfast.

Figure 4-6. Preparing a picket holdfast

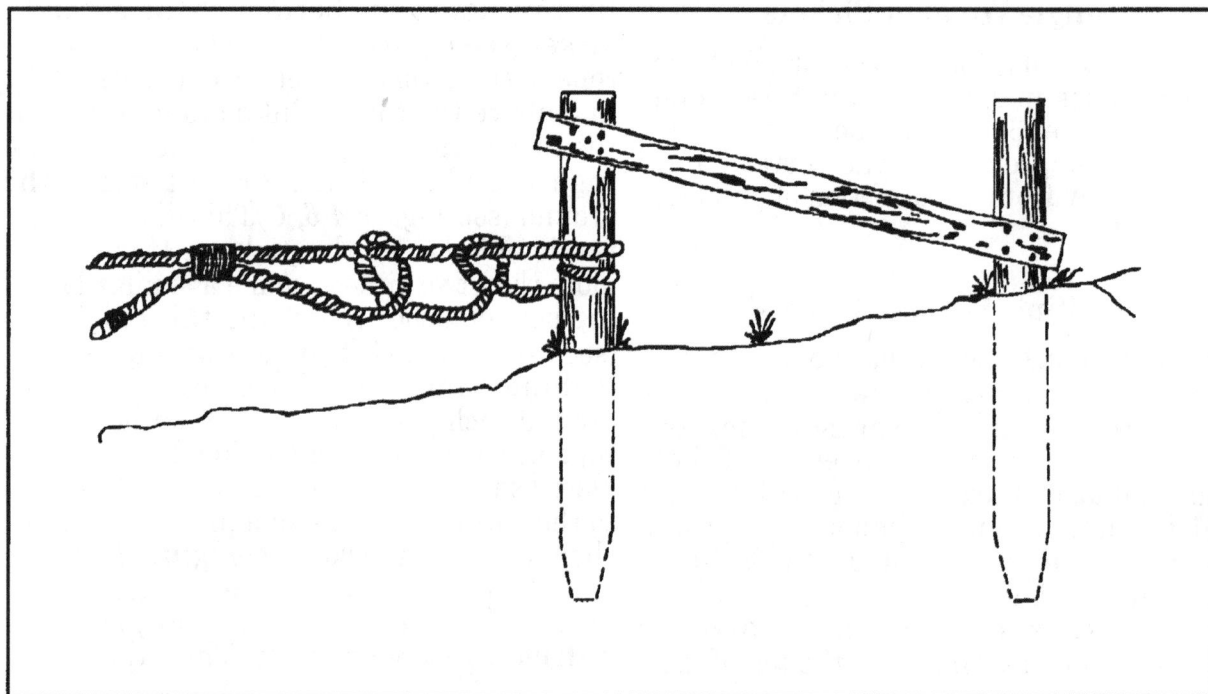

Figure 4-7. Boarded picket holdfast

cipal strength of a multiple-picket holdfast is at the front pickets. Increase the capacity of a holdfast by using two or more pickets to form the front group. This increases both the bearing surface against the soil and the BS.

Steel-Picket Holdfasts

A standard steel-picket holdfast consists of a steel box plate with nine holes drilled through it and a steel eye welded on the end for attaching a guy line (see *Figure 4-8, page 4-6*). The pickets are also steel and are driven through the holes in a way that clinches the pickets in the ground. This holdfast is especially adapted for anchoring horizontal lines, such as the anchor cable on a ponton bridge. Use two or more of these units in combination to provide a stronger anchorage. You can improvise a similar holdfast with a chain by driving steel pickets through the chain links in a crisscross

pattern. Drive the rear pickets in first to secure the end of the chain; then, install the successive pickets so that there is no slack in the chain between the pickets. A lashed steel-picket holdfast consists of steel pickets lashed together with wire rope the same as for a wooden-stake picket holdfast (see *Figure 4-9, page 4-6*). As an expedient, any miscellaneous light-steel members can be driven into the ground and lashed together with wire rope to form an anchorage.

Rock Holdfasts

You can place a holdfast in rock by drilling into the rock and driving the pickets into the holes. Lash the pickets together with a chain (see *Figure 4-10, page 4-7*). Drill the holes about 3 feet apart, in line with the guy line. The first, or front, hole should be 2 1/2 to 3 feet deep and the rear hole, 2 feet deep. Drill the holes at a slight angle, inclined away from the direction of the pull.

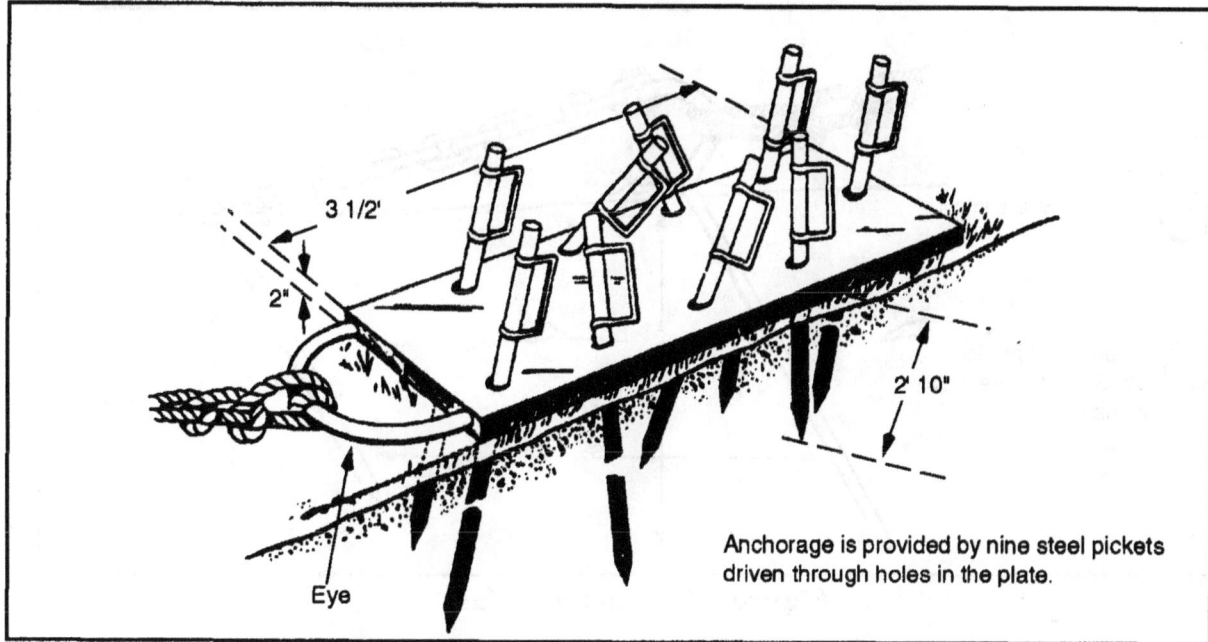

Anchorage is provided by nine steel pickets driven through holes in the plate.

Eye

3 1/2'

2"

2' 10"

Figure 4-8. Standard steel-picket holdfast

COMBINATION HOLDFASTS

For heavy loading of an anchorage, spread the load over the largest possible area of ground. Do this by increasing the number of pickets used. Place four or five multiple picket holdfasts parallel to each other with a heavy log resting against the front pickets to form a combination log and picket holdfast (see *Figure 4-11)*Fasten the guy line or anchor sling to the log that bears against the pickets. The log should bear evenly against all pickets to obtain maximum strength. Select the timber carefully so it can withstand the maximum pull on the line without appreciable bending. Also, you could use a steel cross member to form a combination steel-picket holdfast (see Figure 4-12, page 4-8).

DEADMEN

A deadman is one of the best types of anchorages for heavy loads or permanent installations because of its great holding power.

Construction

You can construct a deadman from a log, a rectangular timber, a steel beam, or a similar object buried in the ground with a guy line or sling attached to its center. This guy line or sling leads to the surface of the ground along a narrow upward sloping trench. The holding power of a deadman is affected by—

• Its frontal bearing area.

• Its mean (average) depth.

Figure 4-9. Lashed steel-picket holdfast

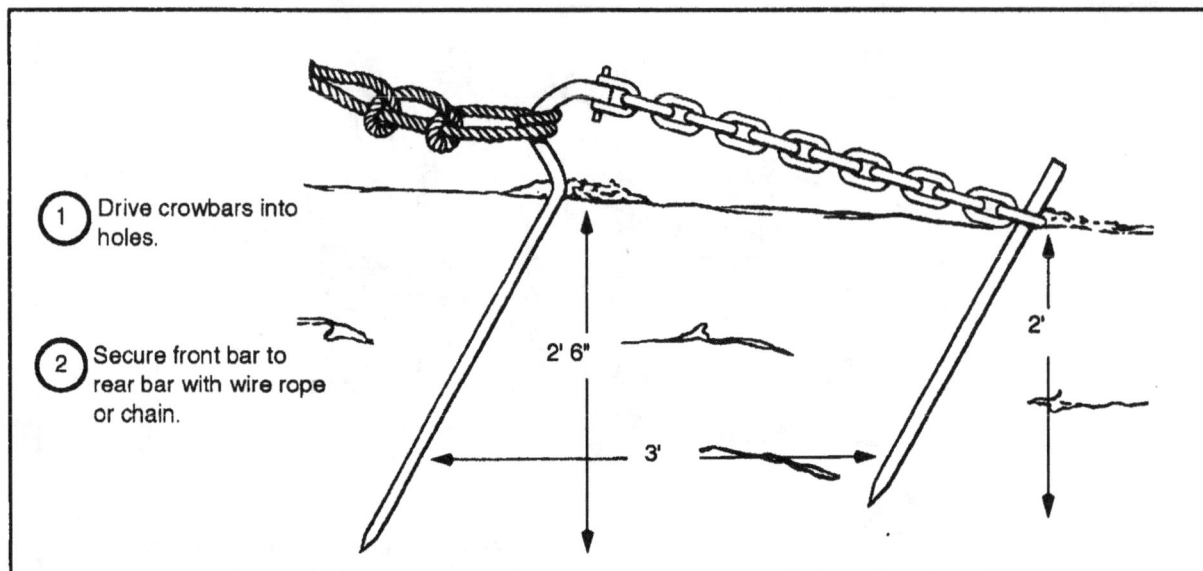

Figure 4-10. Rock holdfast

- The angle of pull.
- The deadman material.
- The soil condition.

The holding power increases progressively as you place the deadman deeper and as the angle of pull approaches a horizontal position (see *Table 4-2, page 4-8*). The holding power of a deadman must be designed to withstand the BS of the line attached to it. In constructing a deadman, dig a hole at right angles to the guy line and undercut 15 degrees from the vertical at the front of the hole facing the load (see *Figure 4-13, page 4-8*). Make the guy line as horizontal as possible, and ensure that the sloping trench matches the slope of the guy line. The main or standing part of the line leads from the bottom of the deadman. This reduces the

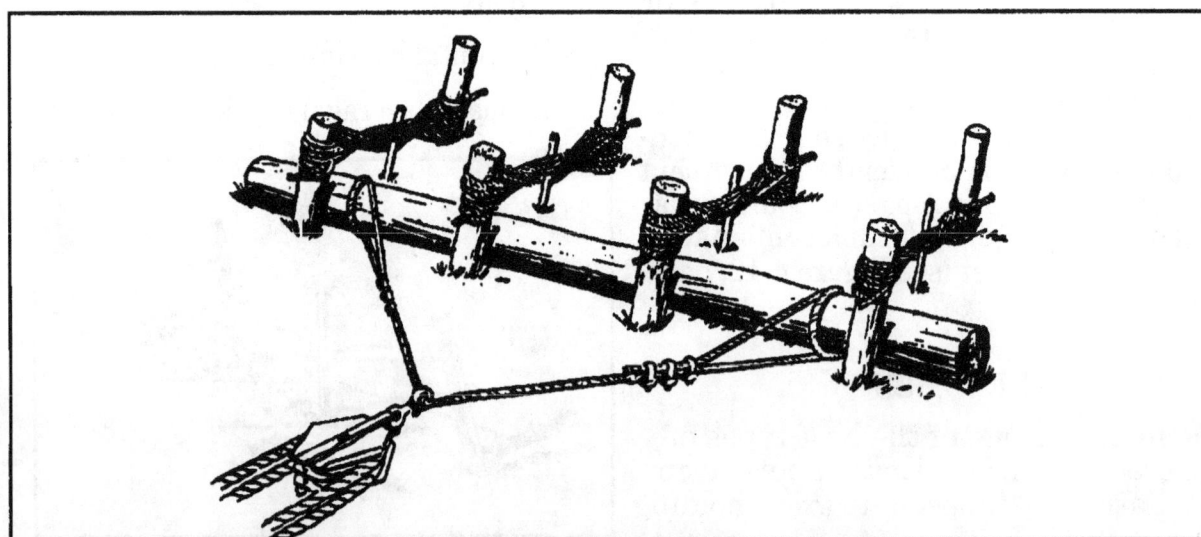

Figure 4-11. Combination log and picket holdfast

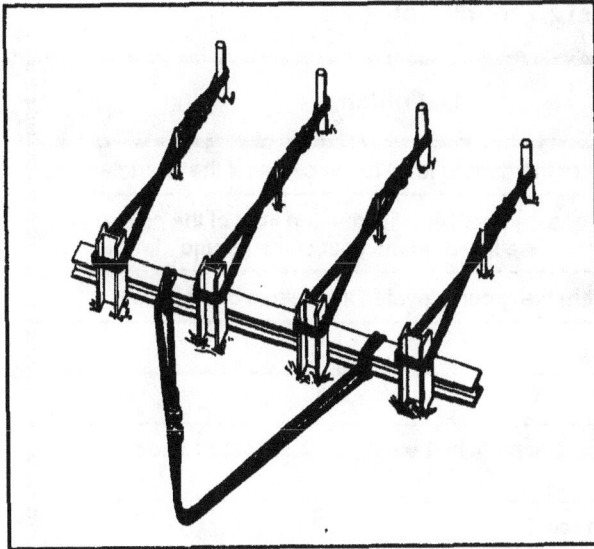

Figure 4-12. Combination steel picket holdfast

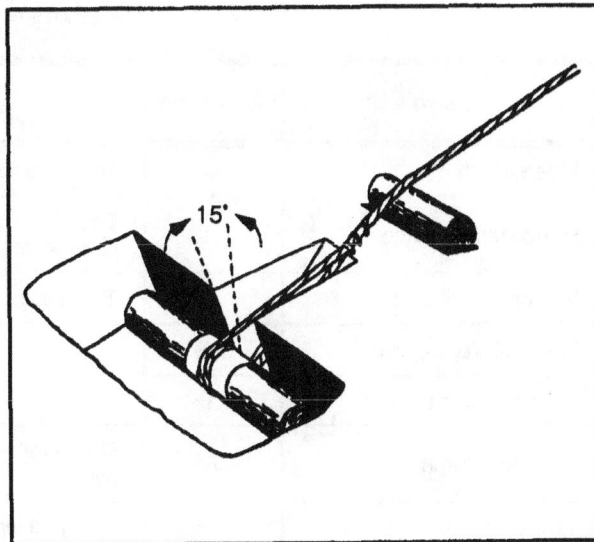

Figure 4-13. Log deadman

tendency to rotate the deadman upward out of the hole. If the line cuts into the ground, place a log or board under the line at the outlet of the sloping trench. When using wire-rope guy lines with a wooden deadman, place a steel bearing plate on the deadman where the wire rope is attached to avoid cutting into the wood. Always place

the wire-rope clips above the ground for retightening and maintenance.

Terms

Table 4-3 lists the terms used in designing a deadman.

Table 4-2. Holding power of deadmen in ordinary soil

Mean Depth of Anchorage (feet)	Inclination of Pull (Vertical to Horizontal) and Safe Resistance of the Projected Area of the Deadman (pounds per square foot [psf])				
	Vertical	1:1 (45°)	1:2 (26.5°)	1:3 (18:5°)	1:4 (14°)
3	600	950	1,300	1,450	1,500
4	1,050	1,750	2,200	2,600	2,700
5	1,700	2,800	3,600	4,000	4,100
6	2,400	3,800	5,100	5,800	6,000
7	3,200	5,100	7,000	8,000	8,400

Table 4-3. Deadman design terminology

Term	Acronym	Definition
Mean depth	MD	The distance from the ground level to the center of the deadman
Horizontal distance	HD	The distance measured horizontally from the front of the hole to the point where the sloping trench comes out of the ground
Vertical depth	VD	The distance from the ground level to the bottom of the hole
Width of sloping trench	WST	
Diameter of timber	D	
Effective length	EL	The length of log that must be bearing against solid or undisturbed soil
Timber length	TL	The total length required
Holding power	HP	The holding power of a deadman in ordinary earth (see *Table 4-2*)
Breaking strength	BS	The breaking strength of rope attached to the deadman
Slope ratio	SR	The slope ratio of the guy line and the sloping trench
Bearing area	BA	The bearing area of the deadman required to hold the BS of the attached rope

Formulas

The following formulas are used in designing a deadman:

- $BA_r = \dfrac{BS}{HP}$

- $EL = \dfrac{BA_r}{D}$

- $TL = EL + WST$

- $VD = MD + \dfrac{D}{2}$

- $HD = \dfrac{VD}{SR}$

A sample problem for designing a deadman is as follows:

Given: l-inch-diameter 6-by-19 IPS rope

$MD = 7\ feet$

$SR = 1.3$

$WST = 2\ feet$

- Requirement I: Determine the length and thickness of a rectangular timber deadman if the height of the face available is 18 inches (1 1/2 feet).

$BS\ of\ wire\ rope = 83,600\ psf\ (see\ Table\ 1\text{-}2)$

$HP = 8,000\ psf\ (see\ Table\ 4\text{-}2)$

Note: Design the deadman so it can withstand a tension equal to the BS of the wire rope

$$BA_r = \frac{BS = 83{,}600 \ pounds}{HP \ 8{,}000 \ psf} = 10.5 \ feet^2$$

$$EL = \frac{BA_r}{face \ height} = \frac{10.5 \ feet^2}{1.5 \ feet} = 7 \ feet$$

$$TL = EL + WST = 7 \ feet + 2 \ feet = 7 \ feet$$

Conduct a final check to ensure that the rectangular timber will not fail by bending by doing a length-to-thickness ratio (L/t), which should be equal to or less than 9. Determine the minimum thickness by L/= 9 and solve for (t):

$$\frac{L}{1 \ t} = 9$$

$$\frac{9}{t} = 9$$

$$= \frac{9}{9} = l \ feet$$

Thus, an 18-inch by 12-inch by 9-foot timber is suitable.

- Requirement II: Determine the length of a log deadman with a diameter of 2 1/2 feet.

$$EL = \frac{BA_r}{D} = \frac{10.5 \ feet^2}{2.5 \ feet} = 4.2 \ feet$$

$$TL = EL + WST = 4.2 \ feet + 2 \ feet = 6.2 \ feet$$

Conduct a final check to ensure that the log will not fail by bending by doing a length-to-diameter ratio (L/d), which should be equal

to or less than 5. The ratio for Requirement II would be equal to L/d = 6.2/2.5 = 2.5. Since this is less than 5, the log will not fail by bending.

Length-to-Diameter Ratio

If the length-to-diameter ratios for a log or a rectangular timber are exceeded, you must decrease the length requirements. Use one of the following methods to accomplish this:

- Increase the mean depth.
- Increase the slope ration (the guy line becomes more horizontal).
- Increase the thickness of the deadman.
- Decrease the width of the sloping trench, if possible.

NOMOGRAPH-DESIGNED DEADMEN

Nomography and charts have been prepared to facilitate the design of deadmen in the field. The deadmen are designed to resist the BS of the cable. The required length and thickness are based on allowable soil bearing with 1-foot lengths added to compensate for the width of the cable trench. The required thickness is based on a L/d ratio of s for logs and a L/d ratio of 9 for cut timber.

Log Deadman

A sample problem for designing a log deadman is as follows:

- Given: 3/4-inch IPS cable. You must bury the required deadman 5 feet at a slope of 1:4.
- Solution: With this information, use the nomograph to determine the diameter and length of the deadman required (see *Figure 4-14*). *Figure 4-15, page 4-12* shows the steps, graphically, on an incomplete nomograph. Lay a straightedge across section A-A

(left-hand scale) on the 5-foot depth at deadman and 1:4 slope and on 3/4-inch IPS on B-B. Read across the straightedge and locate a point on section C-C. Then go horizontally across the graph and intersect the diameter of the log deadmen available. Assume that a 30-inch diameter log is available. Go vertically up from the intersection on the log and read the length of deadman required. In this case, the deadman must be over 5 1/2 feet long. Be careful not to select a log deadman in the darkened area of the nomograph because a log from this area will fail by bending.

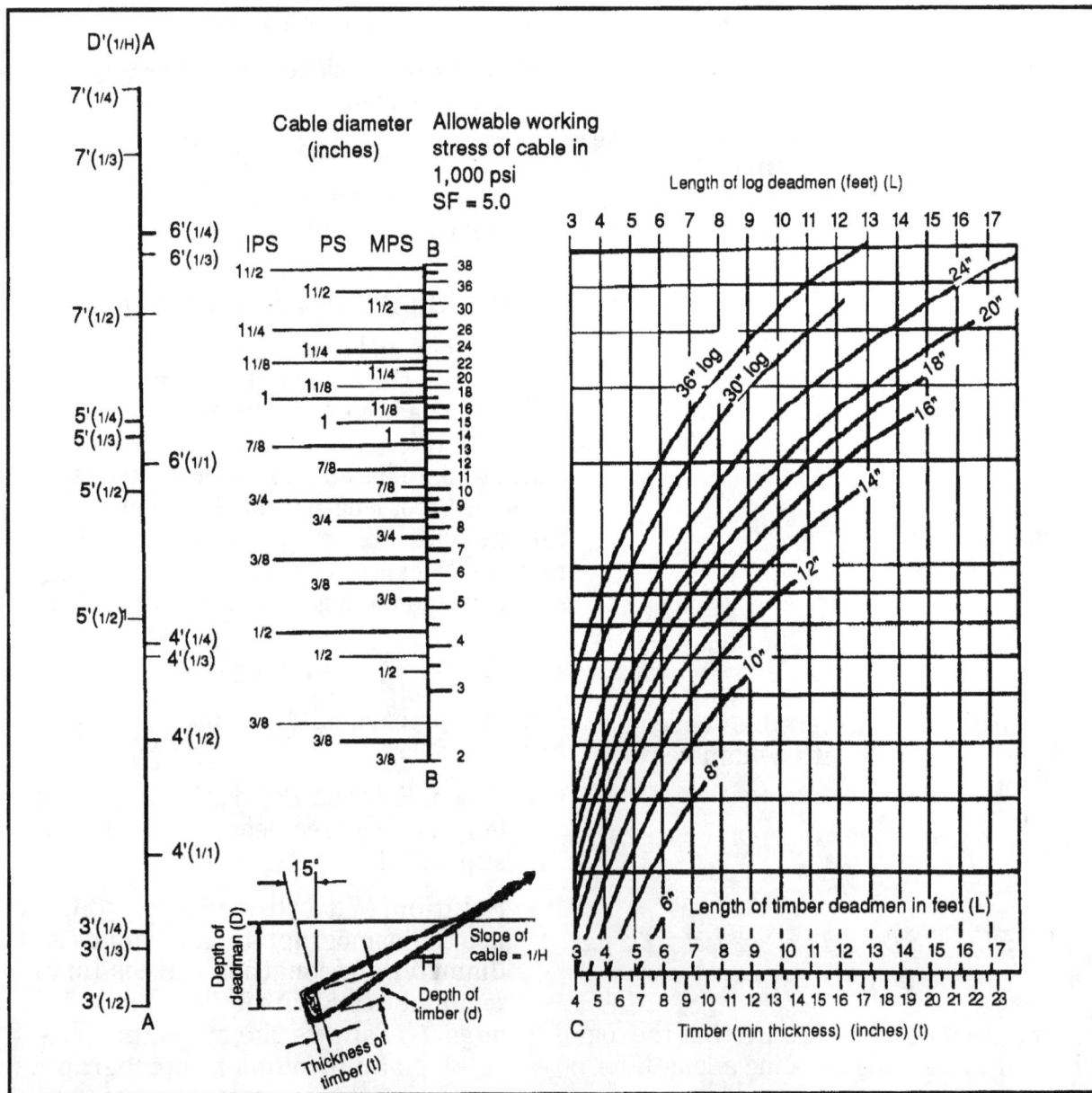

Figure 4-14. Designing a deadman

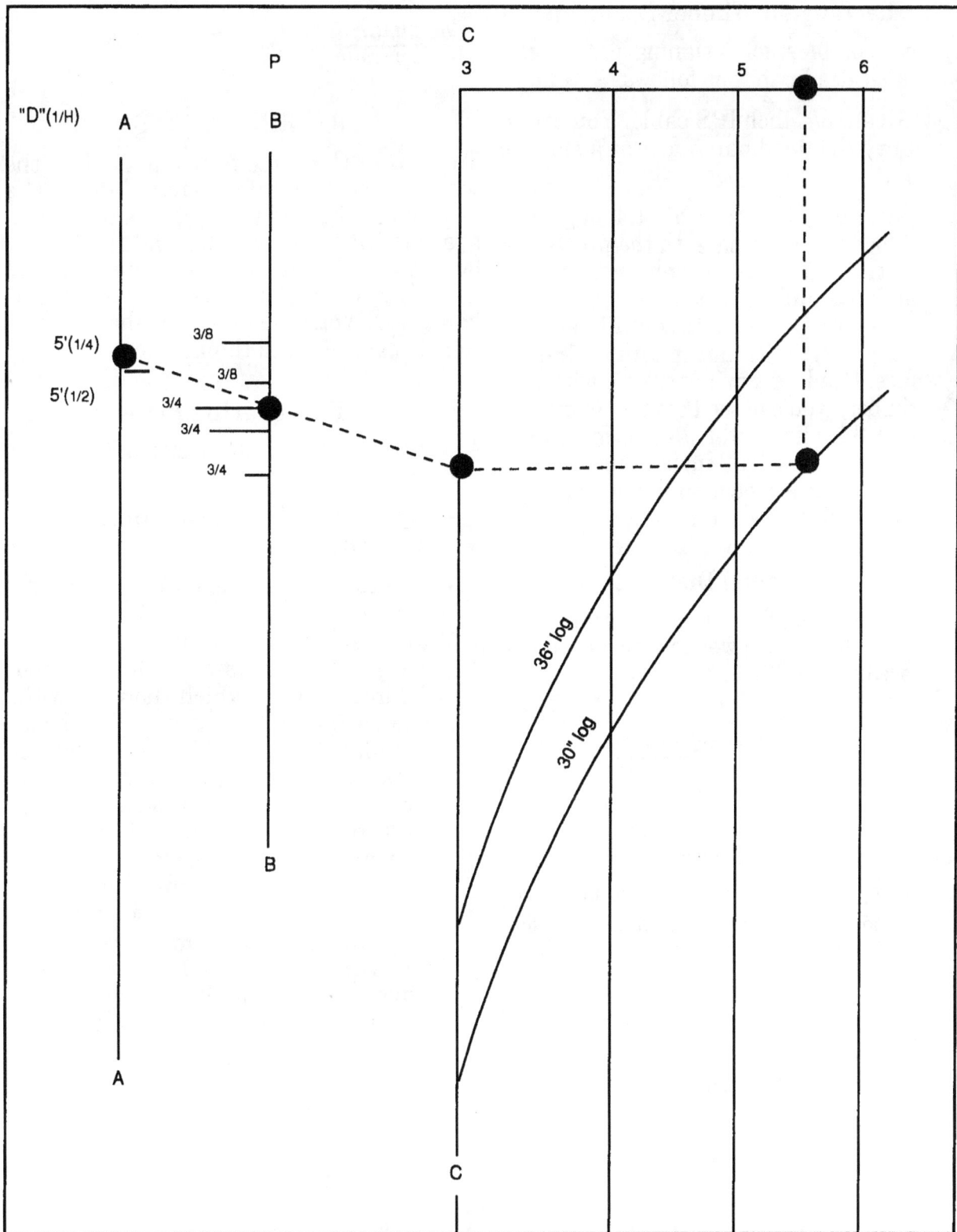

Figure 4-15. Using a nomograph

Rectangular Timber Deadman

A sample problem for designing a rectangular timber deadman is as follows:

- Given: 3/4-inch IPS cable. You are to bury the deadman 5 feet at a slope of 1:4.

- Solution: Use the same 1:4 slope and 5-foot depth, along with the procedure to the left of the graph, as in the previous problem (see Figure 4-14, page 4-11). At C-C, go horizontally across the graph to the timber with an 18-inch face. Reading down (working with cut timber), you can see that the length is 8 feet 6 inches and that the minimum thickness is 11 1/2 inches. None of the timber sizes shown on the nomograph will fail due to bending.

Horizontal Distance

Use the following formula to determine the distance behind the tower in which deadmen are placed:

$$Horizontal\ distance = \frac{tower\ height + deadman\ depth}{slope\ ratio}$$

A sample problem for determining the horizontal distance behind a tower is as follows:

- Given: The tower height is 25 feet 4 1/4 inches, and the deadman depth is 7 feet with a 1:4 slope.

- Solution:

$$\frac{25\ ft\ 4\ 1/4\ in + 7\ ft}{1:4} = \frac{32ft\ 4\ 1/4\ in}{1:4} = 129\ ft\ 5\ in$$

Place the deadman 129 feet behind the tower.

Note: The horizontal distance without a tower is as follows:

$$\frac{deadman\ depth}{slope\ ratio} = \frac{7ft}{1:4} = 28ft$$

BEARING PLATES

To prevent the cable from cutting into the wood, place a metal bearing plate on the deadman. The two types of bearing plates are the flat bearing plate and the formed bearing plate, each with its particular advantages. The flat bearing plate is easily fabricated, while the formed or shaped plate can be made of much thinner steel.

Flat Bearing Plate

A sample problem in the design of flat bearing plates is as follows:

- Given: 12-inch by 12-inch timber 3/4-inch IPS cable

- Solution: Enter the graph (see *Figure 4-16, page 4-14*) from the left of the 3/4-inch cable and go horizontally across the graph to intersect the line marked 12-inch timber, which shows that the plate will be 10 inches wide. (The bearing plate is made 2 inches narrower than the timber to prevent cutting into the anchor cable.) Drop vertically and determine the length of the plate, which is 9 1/2 inches. Go to the top, vertically along the line to where it intersects with 3/4-inch cable, and determine the minimum required thickness, which is 1 1/16 inches. Thus, the necessary bearing plate must be 1 1/16 inches by 9 1/2 inches by 10 inches.

Formed Bearing Plate

The formed bearing plates are either curved to fit logs or formed to fit rectangular timber. In the case of a log, the bearing plate must go half way (180 degrees) around the log. For a shaped timber, the bearing plate

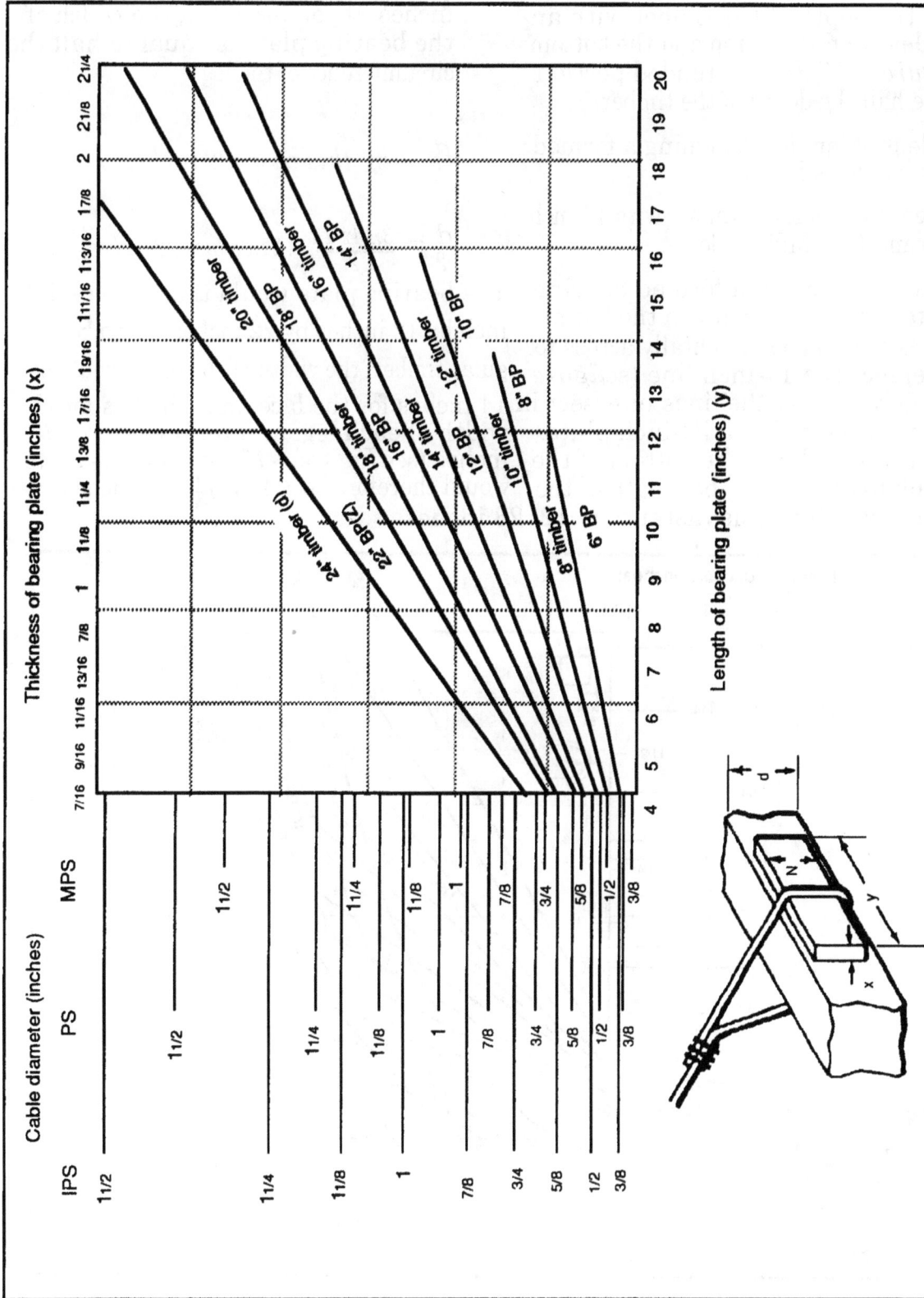

Figure 4-16. Designing a flat bearing plate for a regular deadman

extends the depth of the timber with an extended portion at the top and the bottom (see *Figure 4-17*)Each extended portion should be half the depth of the timber.

A sample problem for designing a formed bearing plate is as follows:

- Given: 14-inch log or timber with 14-inch face and 1 1/8 MPS cable.

- Solution: Design a formed bearing plate. Enter the graph on the left at 1 1/8 MPS and go horizontally across to intersect the 14-inch line (see *figure 4-17*).Note that the lines intersect in an area requiring a l/4-inch plate. Drop vertically to the bottom of the graph to determine the length of the plate, which in this instance is 12

inches. If you use a log, the width of the bearing plate is equal to half the circumference of the log.

$$\frac{d}{2} \quad in \ this \ case, \ 22 \ inches$$

$$\frac{d}{2} = \frac{3.14 \text{x} \ 14}{2} = 21.98 \ (use \ 22 \ inches)$$

The bearing plate would therefore be 1/4 inch by 12 inches by 22 inches. For a rectangular timber, the width of the plate would be 14 inches for the face and 7 inches for the width of each leg, or a total width of 28 inches (see *Figure 4-17*)The bearing plate would therefore be 1/4 inch by 12 inches by 28 inches.

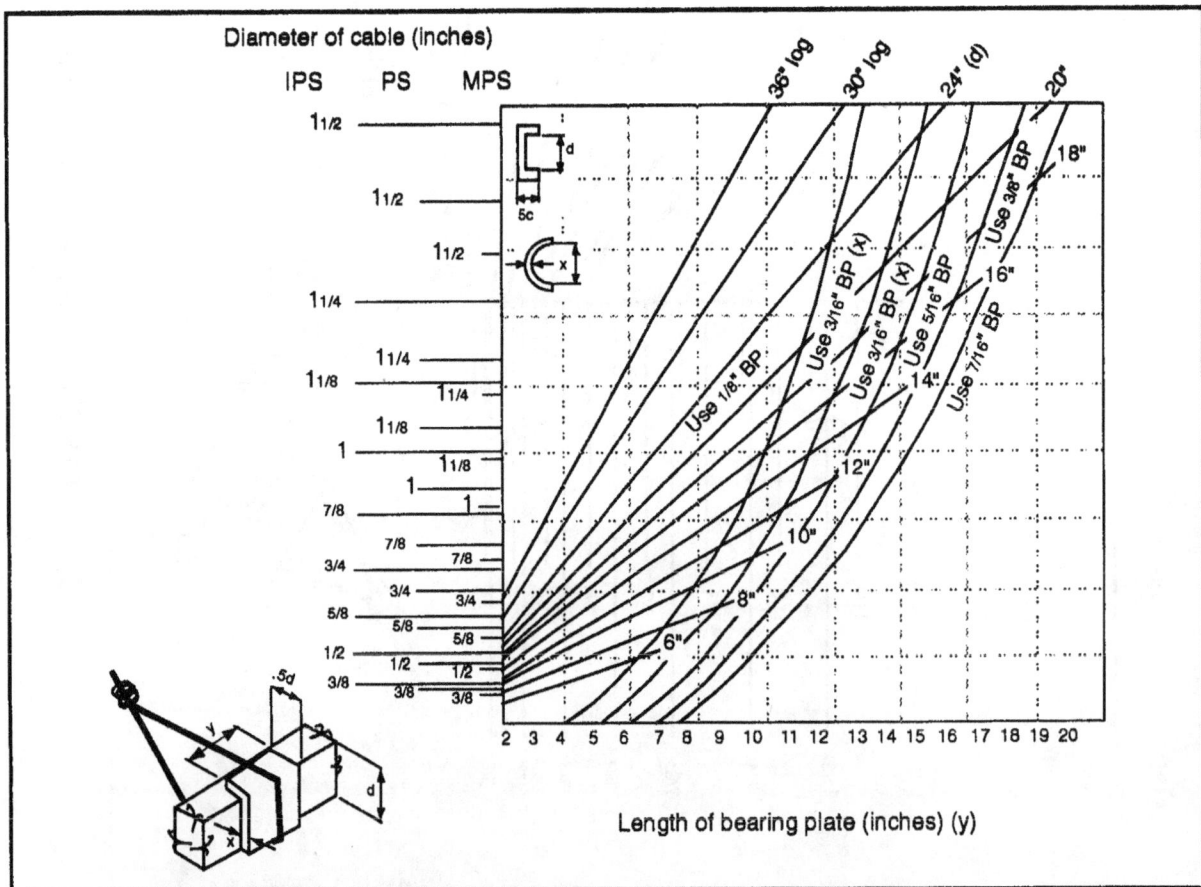

Figure 4-17. Designing a formed bearing plate

Section II. Guy Lines

Guy lines are ropes or chains attached to an object to steady, guide, or secure it. The lines leading from the object or structure are attached to an anchor system (see *Figure 4-18).* When a load is applied to the structure supported by the guy lines, a portion of the load is passed through each supporting guy line to its anchor. The amount of tension on a guy line depends on the—

- Main load.

- Position and weight of the structure.

- Alignment of the guy line with the structure and the main load.

- Angle of the guy line.

For example, if the supported structure is vertical, the stress on each guy line is very small; but if the angle of the structure is 45 degrees, the stress on the guy lines supporting the structure will increase considerably. Wire rope is preferred for guy lines because of its strength and resistance to corrosion. Fiber is also used for guy lines, particularly on temporary structures. The number and size of guy lines required depends on the type of structure to be supported and the tension or pull exerted on the guy lines while the structure is being used.

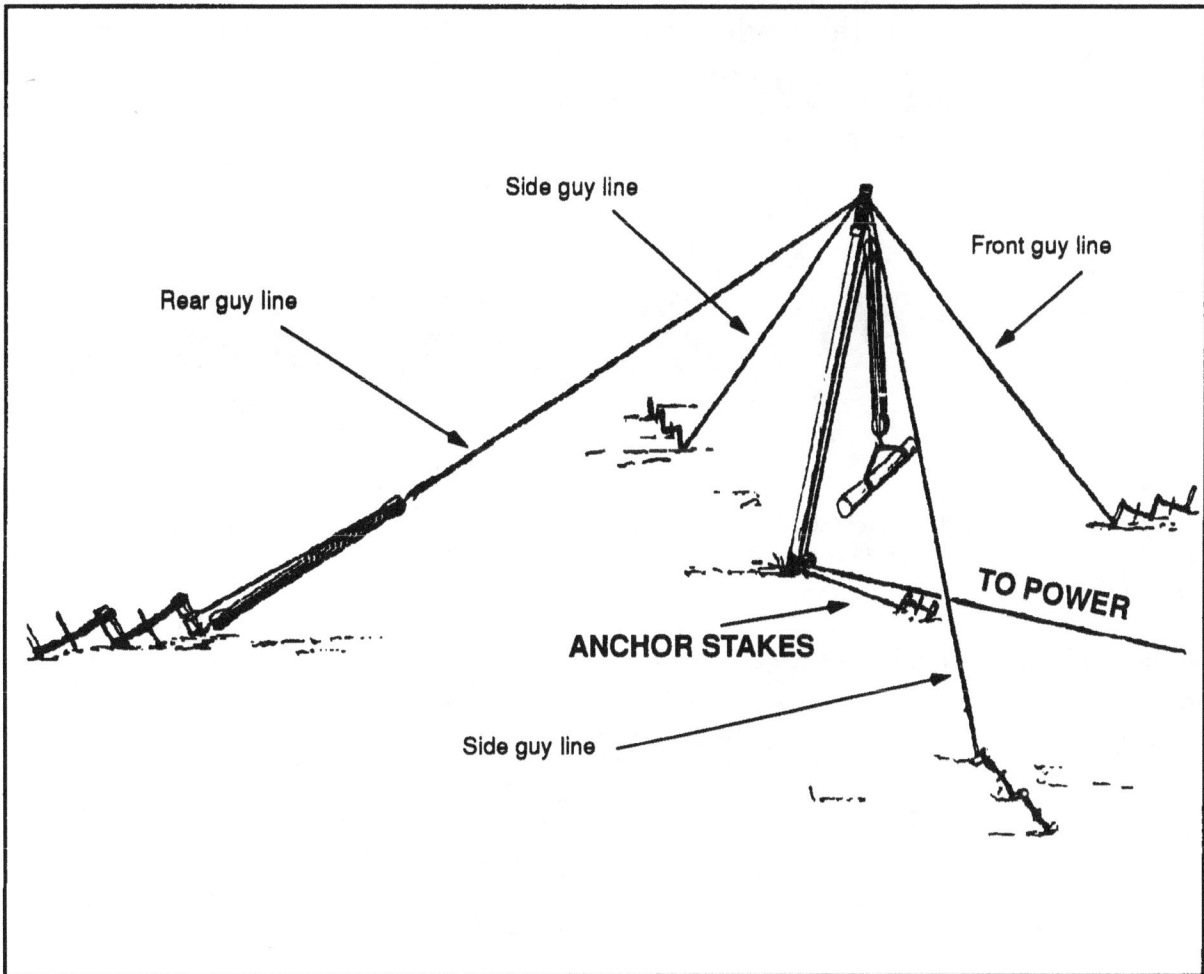

Figure 4-18. Typical guy-line installations

NUMBER OF GUY LINES

Usually a minimum of four guy lines are used for gin poles and boom derricks and two for shears. The guy lines should be evenly spaced around the structure. In a long, slender structure, it is sometimes necessary to provide support at several points in a tiered effect. In such cases, there might be four guy lines from the center of a long pole to anchorage on the ground and four additional guy lines from the top of the pole to anchorage on the ground.

TENSION ON GUY LINES

You must determine the tension that will be exerted on the guy lines beforehand to select the proper size and material you will use. The maximum load or tension on a guy line will result when a guy line is in direct line with the load and the structure. Consider this tension in all strength calculations of guy lines. You can use the following formula to determine the tension for gin poles and shears (see *Figure 4-19, page 4-18*):

$$T = \frac{(W_L + 1/2 W_s) D}{Y}$$

T = Tension in guy line

W_L = *Weight of the load*

W_s = *Weight of spar(s)*

D = *Drift distance, measured from the base of the gin pole or shears to the center of the suspended load along the ground.*

Y= *Perpendicular distance from the rear guy line to the base of the gin pole or, for a shears, to a point on the ground midway between the shear legs.*

A sample problem for determining the tension for gin poles and shears follows:

* Requirement I: gin pole.

Given: W_L = 2,400 lb
W_s = 800 lb
D = 20

Solution:

$$T = \frac{(W_L + 1/2 W_s) D}{Y} = \frac{(2,400 + 1/2\,(800))\,20}{28}$$

= *2,000 pounds of tension in the rear or supporting guy line*

* Requirement II: shears.

Given: The same conditions exist as in Requirement I except that there are two spars, each one weighing 800 pounds.

Solution:

$$T = \frac{(W_L + 1/2 W_s) D)}{Y} = \frac{(2,400 + 1/2\,(800))\,20}{Y}$$

= *2,285 pounds*

NOTE: The shears produced a greater tension in the rear guy line due to the weight of an additional spar.

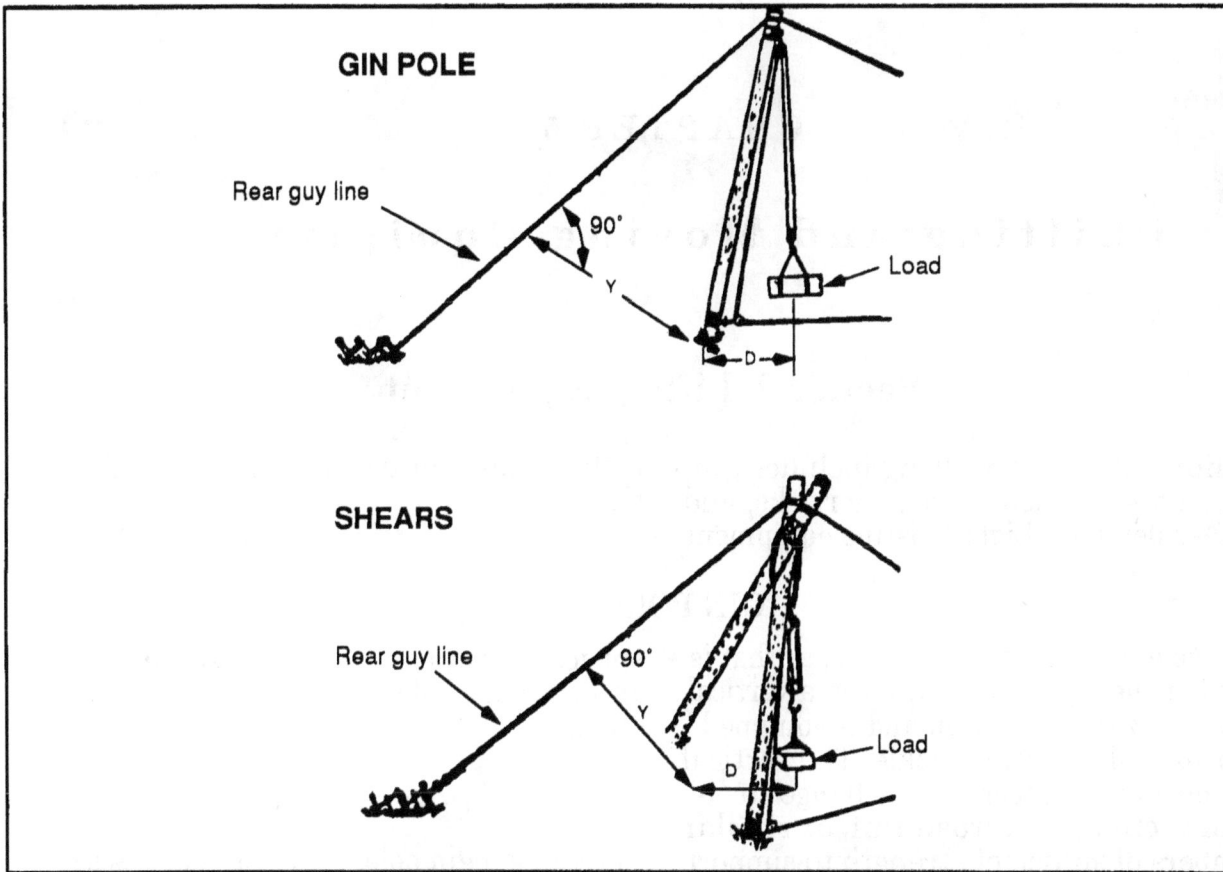

Figure 4-19. Gin pole and shears

SIZE OF GUY LINES

The size of the guy line to use will depend on the amount of tension placed on it. Since the tension on a guy line may be affected by shock loading (and its strength affected by knots, sharp bends, age, and condition), you must incorporate the appropriate FSs. Therefore, choose a rope for the guy line that has a SWC equal to or greater than the tension placed on the guy line.

ANCHORAGE REQUIREMENTS

An ideal anchorage system should be designed to withstand a tension equal to the BS of the guy line attached to it. If you use a 3/8-inch-diameter manila rope as a guy line, the anchorage must be capable of withstanding a tension of 1,350 pounds, which is the BS of the 3/8-inch diameter manila rope. If you use picket holdfasts, you will need at least a 1-1 combination (1,400-pound capacity in ordinary soil). Anchor the guy line as far as possible from the base of the installation to obtain a greater holding power from the anchorage system. The recommended minimum distance from the base of the installation to the anchorage for the guy line is twice the height of the installation.

CHAPTER 5

Lifting and Moving Equipment

Section I. Lifting Equipment

Equipment used for lifting includes gin poles, tripods, shears, boom derricks, and stiff leg derricks. Light hoisting equipment includes pole, brave, and jinniwink derricks.

GIN POLES

A gin pole consists of an upright spar that is guyed at the top to maintain it in a vertical or nearly vertical position and is equipped with suitable hoisting tackle. The vertical spar may be of timber, a wide-flange steel-beam section, a railroad rail, or similar members of sufficient strength to support the load being lifted. The load may be hoisted by hand tackle or by hand- or engine-driven hoists. The gin pole is used widely in erection work because of the ease with which it can be rigged, moved, and operated. It is suitable for raising loads of medium weight to heights of 10 to 50 feet where only a vertical lift is required. The gin pole may also be used to drag loads horizontally toward the base of the pole when preparing for a vertical lift. It cannot be drifted (inclined) more than 45 degrees from the vertical or seven-tenths the height of the pole, nor is it suitable for swinging the load horizontally. The length and thickness of the gin pole depends on the purpose for which it is installed. It should be no longer than 60 times its minimum thickness because of its tendency to buckle under compression. A usable rule is to allow 5 feet of pole for each inch of minimum thickness. *Table 5-1, page 5-2* lists values when using spruce timbers as gin poles, with allowances for normal stresses in hoisting operations.

RIGGING GIN POLES

In rigging a gin pole, lay out the pole with the base at the spot where it is to be erected. To make provisions for the guy lines and tackle blocks, place the gin pole on cribbing for ease of lashing *Figure 4-18, page 4-16* shows the lashing on top of a gin pole and the method of attaching guys. The procedure is as follows:

- Make a tight lashing of eight turns of fiber rope about 1 foot from the top of the pole, with two of the center turns engaging the hook of the upper block of the tackle. Secure the ends of the lashing with a square knot. Nail wooden cleats (boards) to the pole flush with the lower and upper sides of the lashing to prevent the lashing from slipping.

- Lay out guy ropes, each four times the length of the gin pole. In the center of each guy rope, form a clove hitch over

Table 5-1. Safe capacity of spruce timber as gin poles

Size of Timber In (Inches)	Safe Capacity for Given Length of Timber (pounds)					
	20 Feet	25 Feet	30 Feet	40 Feet	50 Feet	60 Feet
6 diameter	5,000	3,000	2,000			
8 diameter		11,000	8,000	5,000	3,000	
10 diameter	31,000	24,000	16,000	9,000	6,000	
12 diameter			31,000	19,000	12,000	9,000
6 x 6	6,000	4,000	3,000			
8 x 8		14,000	10,000	6,000	4,000	
10 x 10	40,000	30,000	20,000	12,000	8,000	
12 x 12			40,000	24,0000	16,000	12,000

Note: Safe capacity of each length of shears or tripod is seven-eights of the value given for a gin pole.

the top of the pole next to the tackle lashing. Be sure to align the guy lines in the direction of their anchors (see *Figure 5-1).*

- Lash a block to the gin pole about 2 feet from the base of the pole, the same as for the tackle lashing at the top, and place a cleat above the lashing to prevent slipping. This block serves as a leading block on the fall line, which allows a directional change of pull from the vertical to the horizontal. A snatch block is the most convenient type to use for this purpose.

- Reeve the hoisting tackle, and use the block lashed to the top of the pole so that the fall line can be passed through the leading block at the base of the gin pole.

- Drive a stake about 3 feet from the base of the gin pole. Tie a rope from the stake to the base of the pole below the lashing on the leading block and near the bottom of the pole. This prevents the pole from skidding while you erect it.

- Check all lines to be sure that they are not snarled. Check all lashings to see that they are made up properly and that all knots are tight. Check the hooks on the blocks to see that they are moused properly. You are now ready to erect the gin pole.

ERECTING GIN POLES

You can easily raise a 40-foot-long gin pole by hand (see *Figure 5-2)*However, you must raise longer poles by supplementary rigging or power equipment. The number of people needed to erect a gin pole depends on the weight of the pole. The procedure is as follows:

- Dig a hole about 2 feet deep for the base of the gin pole.

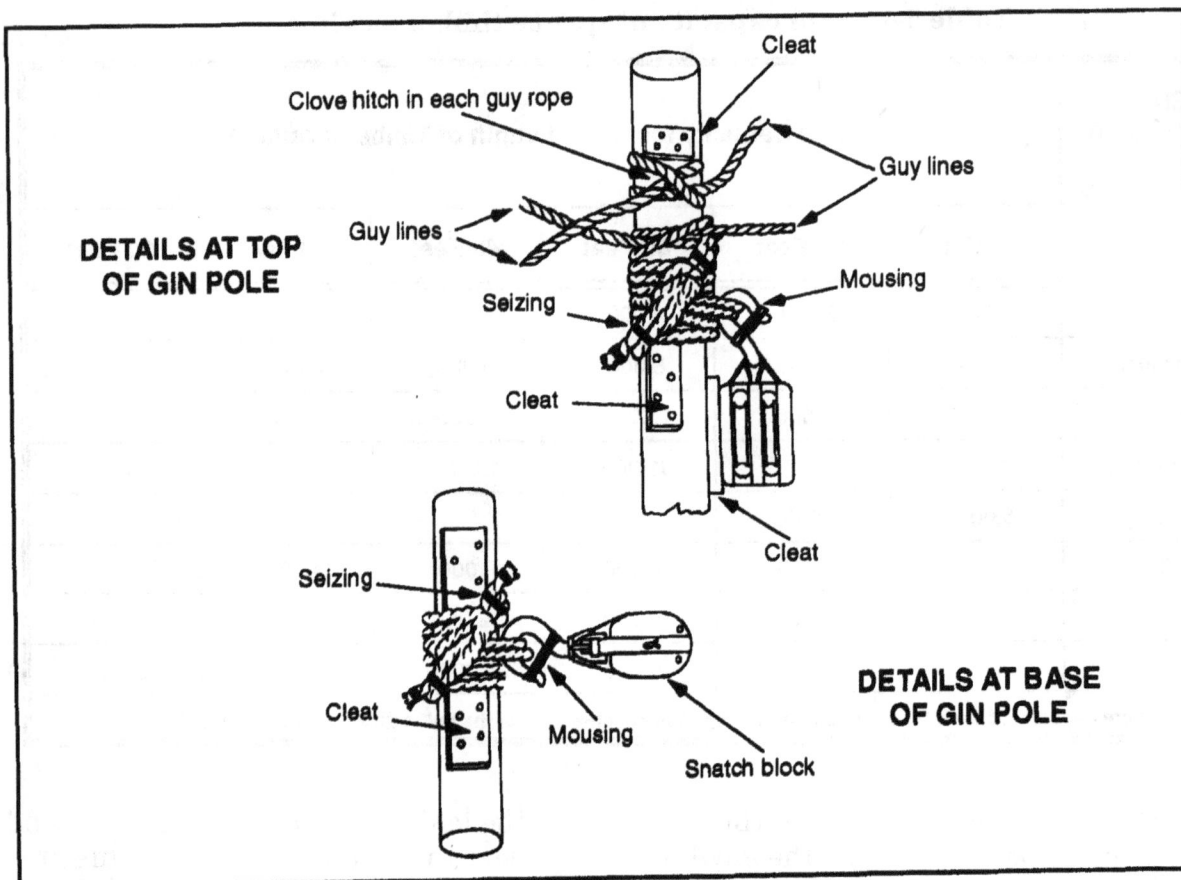

Figure 5-1. Lashing for a gin pole

DETAILS AT TOP OF GIN POLE

Clove hitch in each guy rope

Cleat

Guy lines

Guy lines

Mousing

Seizing

Cleat

Cleat

DETAILS AT BASE OF GIN POLE

Seizing

Cleat

Mousing

Snatch block

Figure 5-2. Erecting a gin pole

- String out the guys to their respective anchorages and assign a person to each anchorage to control the slack in the guy line with a round turn around the anchorage as the pole is raised. If it has not been done already, install an anchorage for the base of the pole.

- Use the tackle system that was used to raise and lower the load to assist in raising the gin pole, if necessary; however, the preferred method is to attach an additional tackle system to the rear guy line. Attach the running block of the rear guy-line tackle system to the rear guy line, the end of which is at this point of erection near the base of the gin pole (see *Figure 4-18, page 4-16*). Secure the fixed or stationary block to the rear anchor. The fall line should come out of the running block to give greater MA to the tackle system. Stretch the tackle system to the base of the gin pole before erecting it to prevent the tackle blocks from chocking.

- Haul in on the fall line of the tackle system, keeping a slight tension on the rear guy line and on each of the side guy lines, while eight people (more for larger poles) raise the top of the pole by hand until the tackle system can take control (see *Figure 5-2, page 5-3*).

- Keep the rear guy line under tension to prevent the pole from swinging and throwing all of its weight on one of the side guys.

- Fasten all guy lines to their anchorages with the round turn and two half hitches when the pole is in its final position, approximately vertical or inclined as desired. At times, you may have to double the portion of rope used for the half hitches.

- Open the leading block at the base of the gin pole and place the fall line from the tackle system through it. When the leading block is closed, the gin pole is ready for use. If you have to drift the top of the pole without moving the base, do it when there is no load on the pole, unless the guys are equipped with tackle.

OPERATING GIN POLES

The gin pole is particularly adapted to vertical lifts (see *Figure 5-3*) Sometimes it is used for lifting and pulling at the same time so that the load being moved travels toward the gin pole just off the ground. When used in this manner, attach a snubbing line of some kind to the other end of the load being dragged; keep it under tension at all times. Use tag lines to control loads that you are lifting vertically. A tag line is a light line fastened to one end of the load and kept under slight tension during hoisting.

TRIPODS

A tripod consists of three legs lashed or secured at the top. The advantage of the tripod over other rigging installations is that it is stable and requires no guy lines to hold it in place. Its disadvantage is that the load can be moved only up and down. The load capacity of a tripod is about one and one-half times that of shears made of the same size material.

RIGGING TRIPODS

The two methods of lashing a tripod, either of which is suitable provided the lashing

Figure 5-3. Hoisting with a gin pole

material is strong enough, are discussed below. The material used for lashing can be fiber rope, wire rope, or chain. Metal rings joined with short chain sections and large enough to slip over the top of the tripod legs also can be used.

Method 1

This method is for fiber rope, 1 inch in diameter or smaller. Since the strength of the tripod is affected directly by the strength of the rope and the lashing used, use more turns than described here for extra heavy loads and fewer turns for light loads. The procedure is as follows:

- Select three spars, about equal in size, and place a mark near the top of each to indicate the center of the lashing.

- Lay two of the spars parallel with their tops resting on a skid or block and a third spar between the first two, with the butt in the opposite direction and the lashing marks on all three in line. The spacing between spars should be about one-half the diameter of the spars. Leave space between the spars so that the lashing will not be drawn too tight when erecting the tripod.

- Make a clove hitch (using a l-inch rope) around one of the outside spars about 4 inches above the lashing mark, and take eight turns of the line around the three spars (see*Figure 5-4, A)*Be sure to maintain the space between the spars while making the turns.

- Finish the lashing by taking two close frapping turns around the lashing between each pair of spars. Secure the end of the rope with a clove hitch on the center spar just above the lashing. Do not draw the frapping turns too tight.

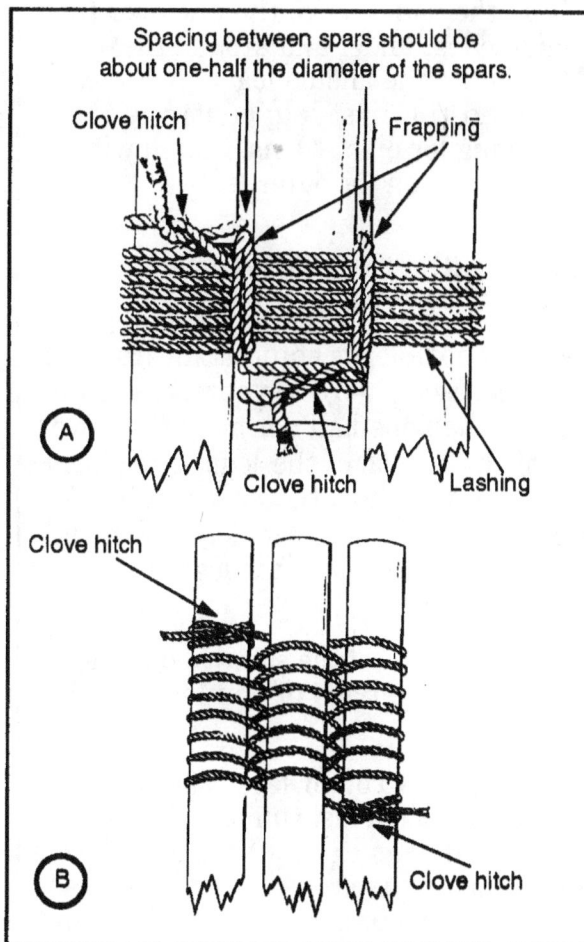

Figure 5-4. Lashing for a tripod

Method H

You can use this method when using slender poles that are not more than 20 feet long or when some means other than hand power is available for erection (see*figure 5-4, B)*The procedure is as follows:

- Lay the three spars parallel to each other with an interval between them slightly greater than twice the diameter of the rope you use. Rest the tops of the poles on a skid so that the ends project over the skid about 2 feet and the butts of the three spars are in line.

- Put a clove hitch on one outside leg at the bottom of the position that the lashing will occupy, which is about 2 feet from the end. Weave the line over the middle leg, under and around the outer leg, under the middle leg, and over and around the first leg; continue this weaving for eight turns. Finish with a clove hitch on the outer leg.

ERECTING TRIPODS

Spread the legs of a tripod in its final position so that each leg is equidistant from the others (see *Figure 5-5)*. This spread should not be less than one-half nor more than two-thirds of the length of the legs. Use chain, rope, or boards to hold the legs in this position. You can lash a leading block for the fall line of the tackle to one of the legs. The procedure is as follows:

- Raise the tops of the spars about 4 feet, keeping the base of the legs on the ground.

- Cross the two outer legs. The third or center leg then rests on top of the cross. With the legs in this position, pass a sling over the cross so that it passes over the top or center leg and around the other two.

- Hook the upper block of a tackle to the sling and mouse the hook.

- Continue raising the tripod by pushing in on the legs as they are lifted at the center. Eight people should be able to raise an ordinary tripod into position.

- Place a rope or chain lashing between the tripod legs to keep them from shifting once they are in their final position.

ERECTING LARGE TRIPODS

For larger tripod installations, you may have to erect a small gin pole to raise the tripod into position. Erect the tripods that are lashed in the manner described in Method II

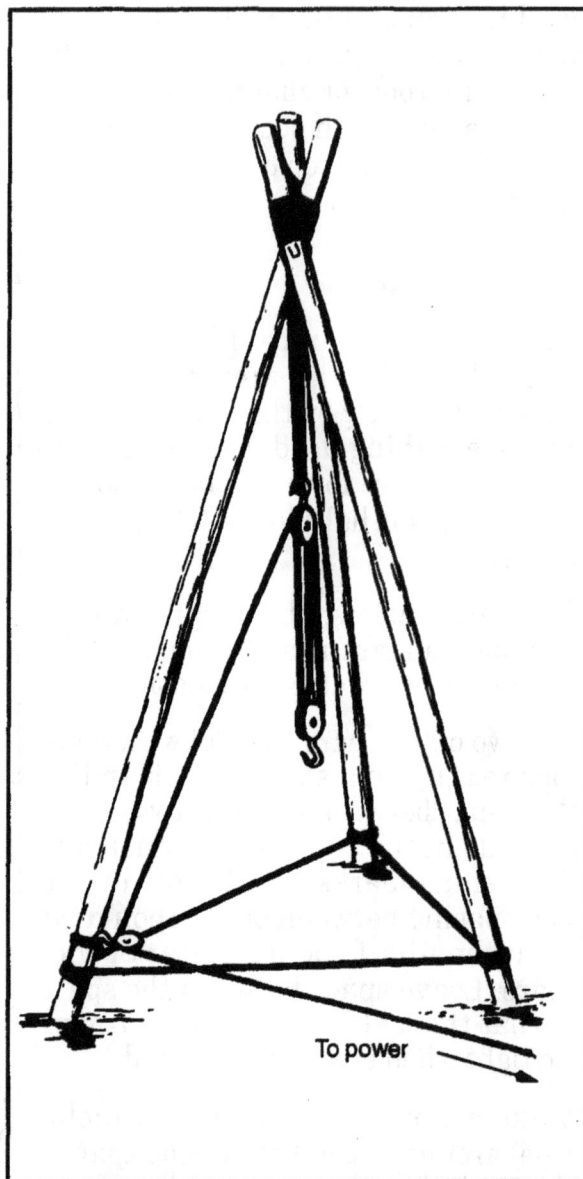

To power

Figure 5-5. Assembled tripod

(with the three legs laid together) by raising the tops of the legs until the legs clear the ground so they can be spread apart. Use guy lines or tag lines to assist in steadying the legs while raising them. Cross the outer legs so that the center leg is on top of the cross, and pass the sling for the hoisting tackle over the center leg and around the two outer legs at the cross.

SHEARS

Shears made by lashing two legs together with a rope are well adapted for lifting heavy machinery or other bulky loads. They are formed by two members crossed at their tops, with the hoisting tackle suspended from the intersection. Shears must be guyed to hold them in position. Shears are quickly assembled and erected. They require only two guys and are adapted to working at an inclination from the vertical. The legs of the shears may be round poles, timbers, heavy planks, or steel bars, depending on the material at hand and the purpose of the shears. In determining the size of the members to use, the load to be lifted and the ratio ($L/_d$) of the legs are the determining factors. For heavy loads, the L_d should not exceed 60 because of the tendency of the legs to bend rather than to act as columns. For light work, you can improvise shears from two planks or light poles bolted together and reinforced by a small lashing at the intersection of the legs.

RIGGING SHEARS

When the shears are erected, the spread of the legs should equal about one-half the height of the shears. The maximum allowable drift is 45 degrees. Tackle blocks and guys for shears are essential. You can secure the guy ropes to firm posts or trees with a turn of the rope so that the length of the guys can be adjusted easily. The procedure is as follows:

- Lay two timbers together on the ground in line with the guys, with the butt ends pointing toward the back guy and close to the point of erection.

- Place a large block under the tops of the legs just below the point of lashing and insert a small spacer block between the tops at the same point (see *Figure 5-6*). The separation between

the legs at this point should be equal to one-third the diameter of one leg to make handling of the lashing easier.

- With sufficient l-inch rope for 14 turns around both legs, make a clove hitch around one spar and take eight turns around both legs above the clove hitch (see *Figure 5-6*). Wrap the turns tightly so that the lashing is smooth and without kinks.

- Finish the lashing by taking two frapping turns around the lashing between the legs and securing the end of the rope to the other leg just below the lashing. For handling heavy loads, increase the number of lashing turns.

ERECTING SHEARS

Dig the holes at the points where the legs of the shears are to stand. If placed on rocky ground, make sure that the base for the shears is level. Cross the legs of the shears and place the butts at the edges of the holes. With a short length of rope, make two turns over the cross at the top of the shears and tie the rope together to form a sling. Be sure to have the sling bearing against the spars and not on the shears lashing entirely. The procedure is as follows:

- Reeve a set of blocks and place the hook of the upper block through the sling. Secure the sling in the hook by mousing. Fasten the lower block to one of the legs near the butt so that it will be in a convenient position when the shears have been raised but will be out of the way during erection.

- Rig another tackle in the back guy near its anchorage if you use the shears on heavy lifts. Secure the two guys to the

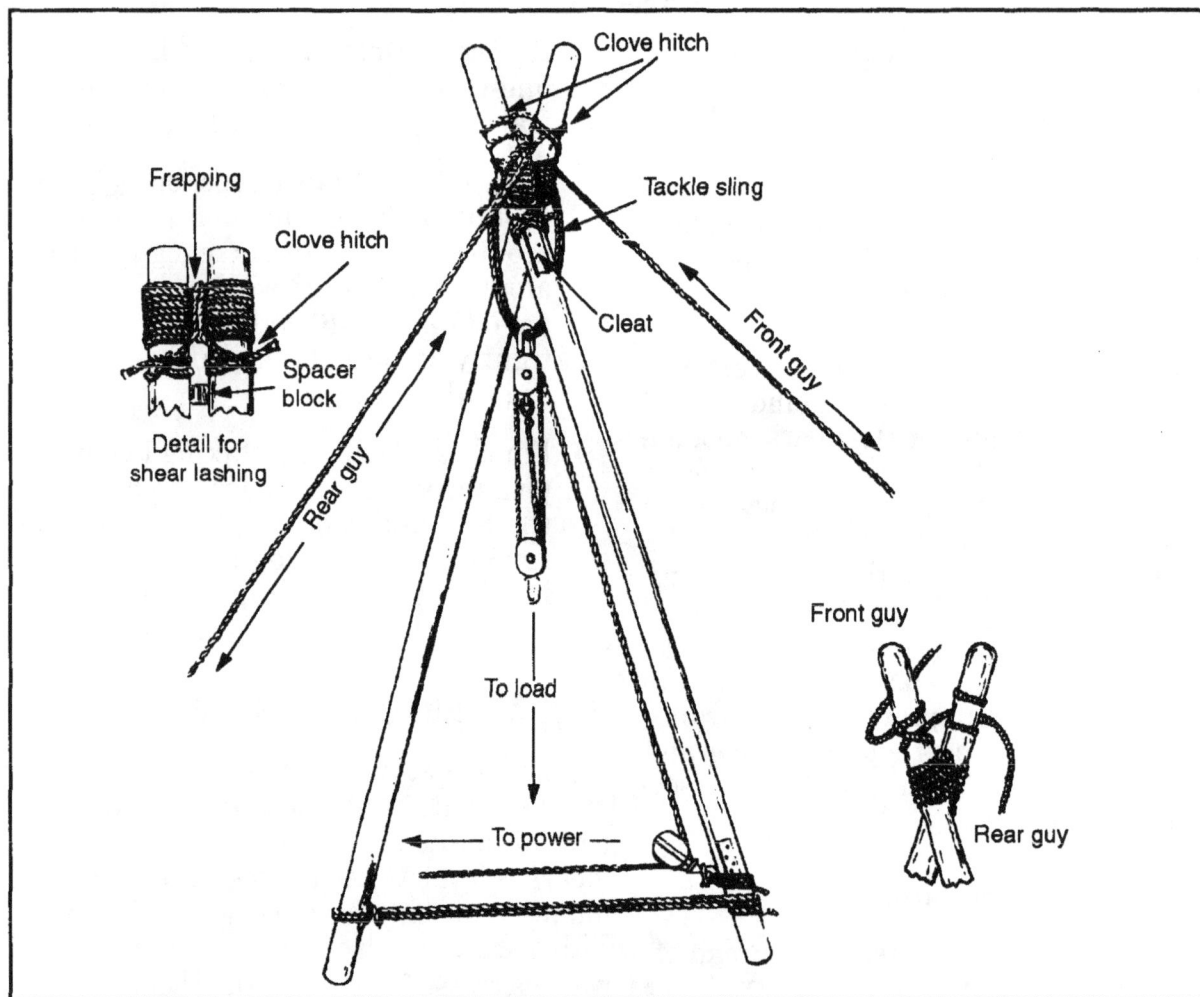

Figure 5-6. Lashing for shears

top of the shears with clove hitches to legs opposite their anchorages above the lashing.

• Lift the top end of the shears legs and "walk" them up by hand until the tackle on the rear guy line can take effect (see *Figure 5-7, page 5-10*). It will take several people (depending on the size of the shears) to do this. Then raise the shears legs into final position by hauling in on the tackle. Secure the front guy line to its anchorage before raising the shears legs, and keep a slight tension on this line to control movement.

• Keep the legs from spreading by connecting them with rope, a chair, or boards. It may be neceesary, under some conditions, to anchor each leg of the shears while erecting them to keep the legs from sliding in the wrong direction.

OPERATING SHEARS

The rear guy is a very important part of the shears rigging, since it is under a considerable strain during hoisting. To avoid guy-line failure, design them according to the principles discussed in *Chapter 4, Section II*. The front guy has very little strain on it and is used mainly to aid in adjusting the drift and to steady the top of the shears when hoisting or placing the load. You may have to rig a tackle in the rear guy for handling heavy loads. During operation, set the desired drift by adjusting the rear guy, but do not do this while a load is on-the shears. For handling light loads, the fall line of the tackle of the shears can be led straight out of the upper block. When handling heavy loads, you may have to lash a snatch block near the base of one of the shear legs to act as a leading block (see *Figure 5-8*)Run the fall line through the leading block to a hand- or power-operated winch for heavy loads.

Figure 5-7. Erecting shears

Figure 5-8. Hoisting with shears

BOOM DERRICKS

A boom derrick is a lifting device that incorporates the advantages of a gin pole and the long horizontal reach of a boom. Use the boom derrick to lift and swing medium-size loads in a 90-degree arc on either side of the resting position of the boom, for a total swing of 180 degrees. When employing a boom derrick in lifting heavy loads, set it on a turn plate or turn wheel to allow the mast and boom to swing as a unit. A mast is a gin pole used with a boom. The mast can swing more than 180 degrees when it is set on a turn plate or turn wheel.

RIGGING BOOM DERRICKS

For hoisting medium loads, rig a boom to swing independently of the pole. Take care to ensure the safety of those using the installation. Use a boom only temporarily or when time does not permit a more stable installation. When using a boom on a gin pole, more stress is placed on the rear guy; therefore, you may need a stronger guy. In case larger rope is not at hand, use a set of tackle reeved with the same size rope as that used in the hoisting tackle as a guy line by extending the tackle from the top of the gin pole to the anchorage. Lash the block attached to the gin pole at the point where the other guys are tied and in the same manner. The procedure is as follows:

- Rig a gin pole as described on *page 5-1*, but lash another block about 2 feet below the tackle lashing at the top of the pole (see *Figure 5-9*). Reeve the tackle so that the fall line comes from the traveling block instead of the standing block. Attach the traveling block to the top end of the boom after erecting the gin pole.

- Erect the gin pole in the manner described on *page 5-1*, but pass the fall line of the tackle through the extra block at the top of the pole before erecting it to increase the MA of the tackle system.

- Select a boom with the same diameter and not more than two-thirds as long as the gin pole. Spike two boards to the butt end of the boom and lash them with rope, making a fork (see *Figure 5-9*). Make the lashing with a minimum of sixteen turns and tie it off with a square knot. Drive wedges under the lashing next to the cleats to help make the fork more secure (see *Figure 5-9*).

- Spike cleats to the mast about 4 feet above the resting place of the boom and place another block lashing just above these cleats. This block lashing will support the butt of the boom. If a separate tackle system is rigged up to support the butt of the boom, place an additional block lashing on the boom just below the larger lashing to secure the running block of the tackle system.

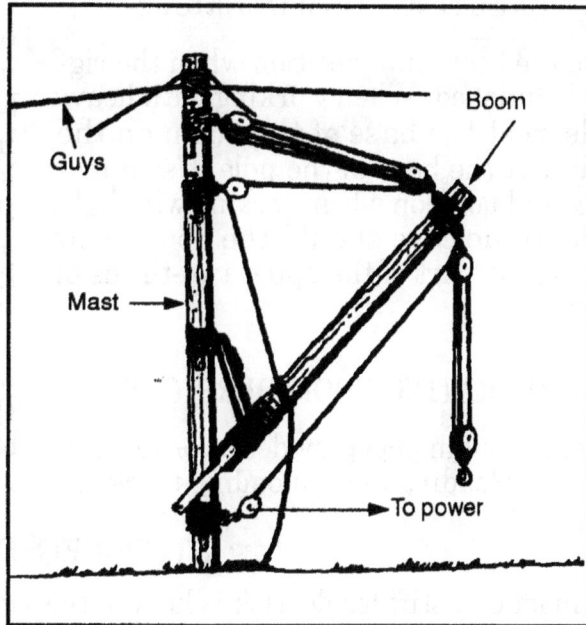

Figure 5-9. Rigging a boom on a gin pole

- Use manpower to lift the boom in place on the mast through the sling that will support it if the boom is light enough. The sling consists of two turns of rope with the ends tied together with a square knot. The sling should pass through the center four turns of the block lashing on the mast and should cradle the boom. On heavier booms, use the tackle system on the top of the mast to raise the butt of the boom to the desired position onto the mast.

- Lash the traveling block of the gin pole tackle to the top end of the boom as described on *page 5-1* and lash the standing block of the boom tackle at the same point. Reeve the boom tackle so that the fall line comes from the standing block and passes through the block at the base of the gin pole. The use of the leading block on this fall line is optional, but when handling heavy loads, apply more power to a horizontal line leading from the block with less strain on the boom and guys.

ERECTING BOOM DERRICKS

Raise the boom into position when the rigging is finished. When working with heavy loads, rest the base of the boom on the ground at the base of the pole. Use a more horizontal position when working with light loads. In no case should the boom bear against any part of the upper two-thirds of the mast.

OPERATING BOOM DERRICKS

A boom on a gin pole provides a convenient means for loading and unloading trucks or flatcars when the base of the gin pole cannot be set close to the object to be lifted. It is used also on docks and piers for unloading boats and barges. Swing the boom by pushing directly on the load or by pulling the load with bridle lines or tag lines. Adjust the angle of the boom to the mast by hauling on the fall line of the mast tackle. Raise or lower the load by hauling on the fall line of the boom tackle. You should place a leading block (snatch block) at the base of the gin pole. Lead the fall line of the boom tackle through this leadingblock to a hand- or power-operated winch for the actual hoisting of the load.

STIFF-LEG DERRICKS

The mast of a stiff-leg derrick is held in the vertical position by two rigid, inclined struts connected to the top of the mast. The struts are spread 60 to 90 degrees to provide support in two directions and are attached to sills extending from the bottom of the mast. The mast is mounted on vertical pins. The mast and boom can swing through an arc of about 270 degrees. The tackles for hoisting the load and raising the boom are similar to those used with the boom and gin pole (see *page 5-11, Rigging Boom Derrick*).

mounted on the tower. The stiff-leg derrick also is used where guy lines cannot be provided, as on the edge of a wharf or on a barge.

STEEL DERRICKS

Steel derricks of the stiff-leg type are available to engineer troops in two sizes:

- A 4-ton rated capacity with a 28-foot radius (see *Figure 5-10, page 5-14*).

- A 30-ton rated capacity with a 38-foot radius, when properly counterweighted.

Both derricks are erected on fixed bases. The 4-ton derrick, including a skid-mounted, double-drum, gasoline-engine-driven hoist, weighs 7 tons and occupies a space 20 feet square. The 30-ton derrick, including a skid-mounted, double-drum hoist, weighs about 22 tons and occupies a space 29 feet square.

OPERATING STIFF-LEG DERRICKS

A stiff-leg derrick equipped with a long boom is suitable for yard use for unloading and transferring material whenever continuous operations are carried on within reach of its boom. When used on a bridge deck, move these derricks on rollers. They are sometimes used in multistory buildings surmounted by towers to hoist material to the roof of the main building to supply guy derricks

LIGHT HOISTING EQUIPMENT

Extended construction projects usually involve erecting numerous light members as well as the heavy main members. Progress can be more rapid if you raise the light members by hand or by light hoisting equipment, allowing the heavy hoisting equipment to move ahead with the erection of the main members. Very light members can be

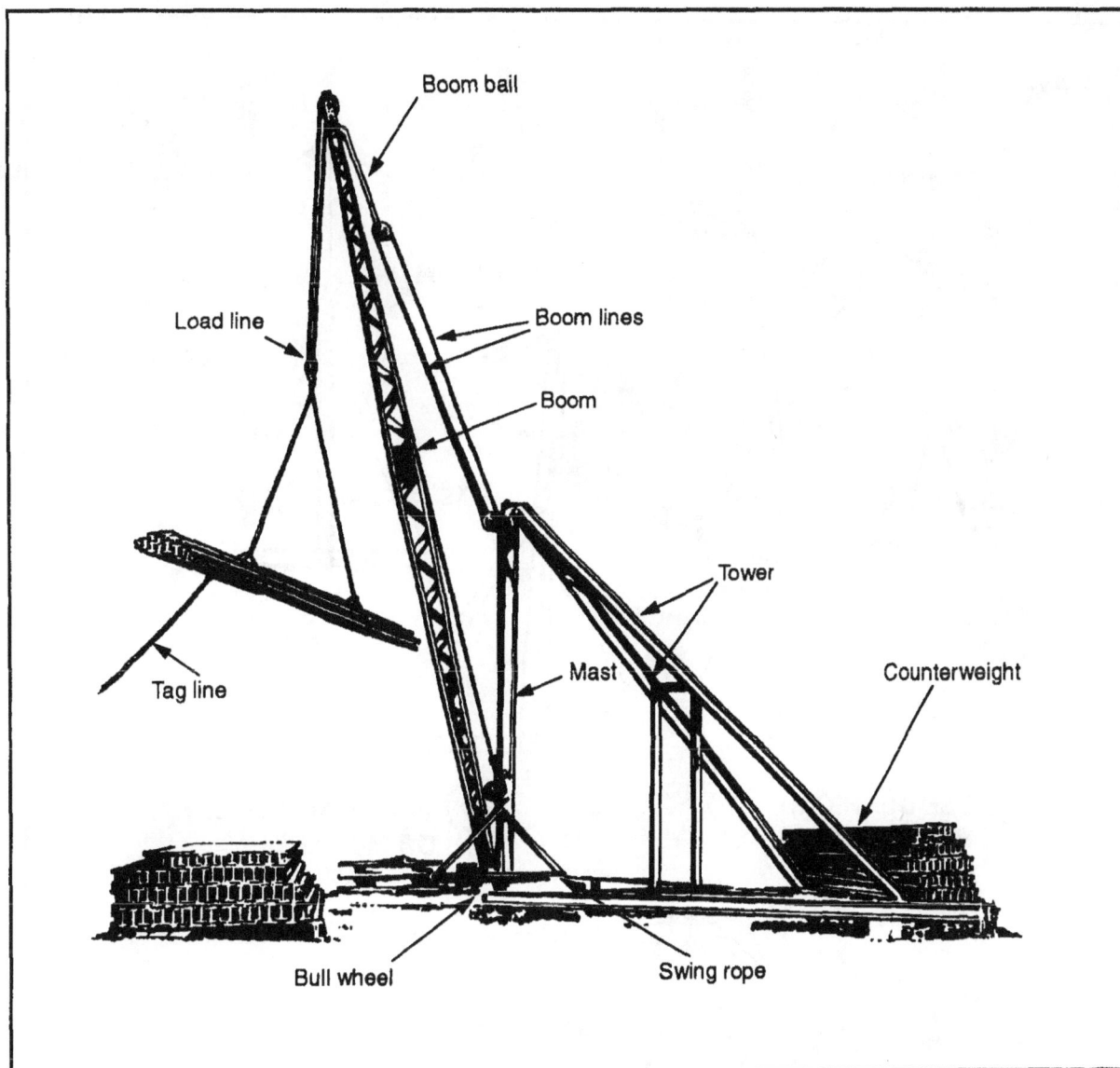

Figure 5-10. Four-ton stiff-leg derrick

raised into place by two people using manila handlines. When handlines are inadequate or when members must be raised above the working level, use light hoisting equipment. Many types of hoisting equipment for lifting light loads have been devised. Those discussed here are only typical examples that can be constructed easily in the field and moved readily about the job.

POLE DERRICKS

The improved pole derrick, called a "dutchman", is essentially a gin pole constructed with a sill and knee braces at the bottom (see *Figure 5-11, A*). It is usually installed with guys at the front and back. It is effective for lifting loads of 2 tons and, because of its light weight and few guys, is readily moved from place to place by a small squad.

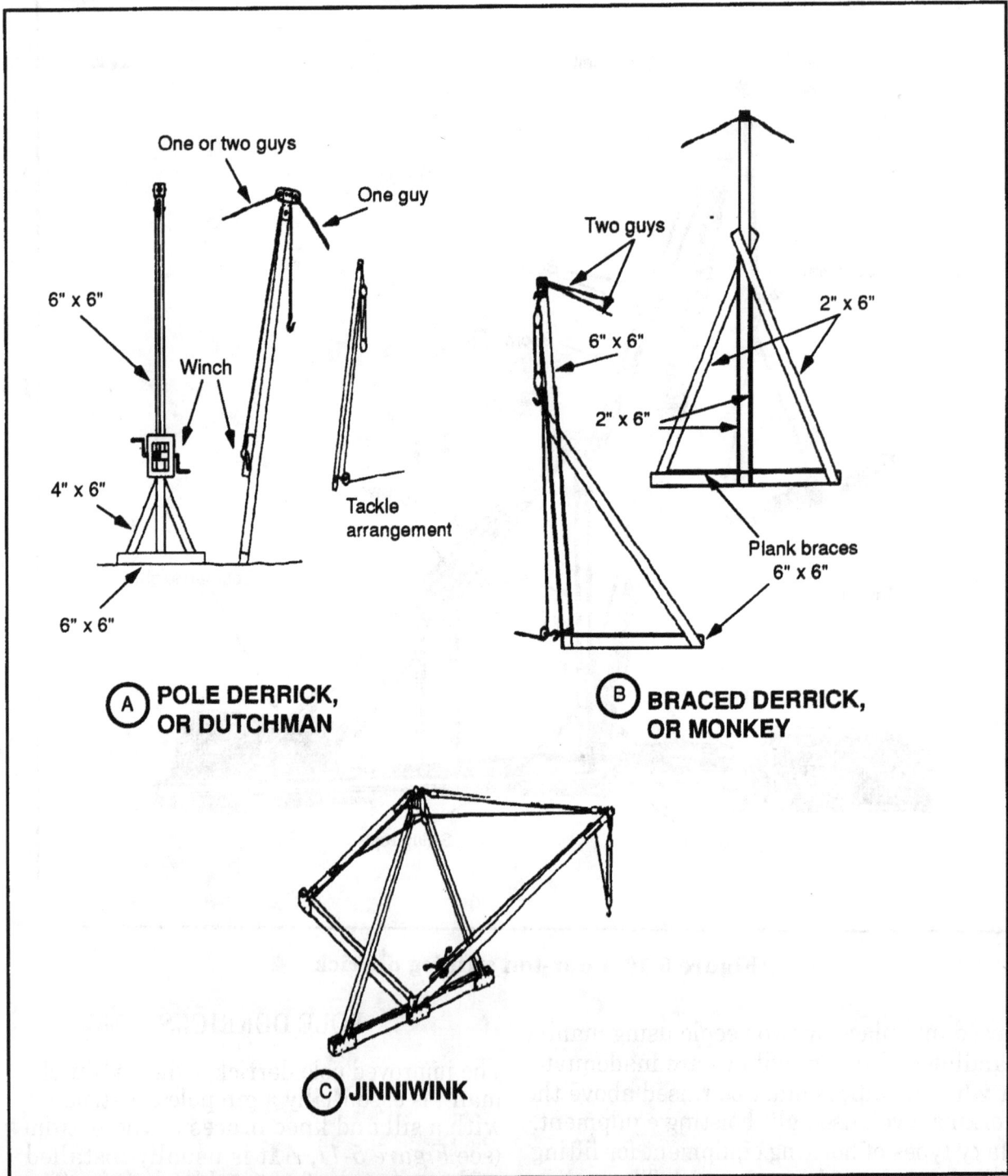

Figure 5-11. Light hoisting equipment

BRAVE DERRICKS

The braced derrick, known as a "monkey", is very useful for filling in heavy members behind the regular erection equipment (see *Figure 5-11, B, page 5-15*). Two back guys are usually employed when lifting heavy loads, although light members may be lifted without them. Power is furnished by a hand- or power-driven hoist. The construction of the base of the monkey permits it to be anchored to the structure by lashings to resist the pull of the lead line on the snatch block at the foot of the mast.

JINNIWINK DERRICKS

This derrick is suitable for lifting loads weighing 5 tons (see *Figure 5-11, C, page 5-15*). Hand-powered jinniwinks are rigged preferably with manila rope. Those operated by a power-driven hoist should be rigged with wire rope, The jinniwink is lashed down to the structural frame at both the front sill and tail sill to prevent the tail sill from rising when a load is lifted.

Section II. Moving Equipment

Skids, rollers and jacks are used to move heavy loads. Cribbing or blocking is often necessary as a safety measure to keep an object in position or to prevent accidents to people who work under or near these heavy objects. Cribbing is formed by piling timbers in tiers, with the tiers alternating in direction, to support a heavy weight at a height greater than blocking would provide (see *Figure 5-12*). A firm and level foundation for cribbing is essential, and the bottom timbers should rest firmly and evenly on the ground. Blocking used as a foundation for jacks should be sound and large enough to carry the load. The timbers should be dry, free from grease, and placed firmly on the ground so that the pressure is evenly distributed.

SKIDS

Place timber skids longitudinally under heavy loads either to—

- Distribute the weight over a greater area.

- Make a smooth surface for skidding equipment.

- Provide a runway surface when rollers are used (see *Figure 5-13*).

Oak planks 2 inches thick and about 15 feet long make satisfactory skids for most operations. Keep the angle of the skids low to prevent the load from drifting or getting out of control. You can use grease on skids when only horizontal movement is involved; however, in most circumstances, greasing is dangerous because it may cause the load to drift sideways suddenly.

ROLLERS

Use hardwood or pipe rollers over skids for moving very heavy loads into position. Place the skids under the rollers to provide a smooth, continuous surface for the rollers. Make sure that the rollers are smooth and round and long enough to pass completely under the load being moved. Support the load on longitudinal wooden members to provide a smooth upper surface for the rollers to move on. The skids placed underneath

Figure 5-12. Timber cribbing

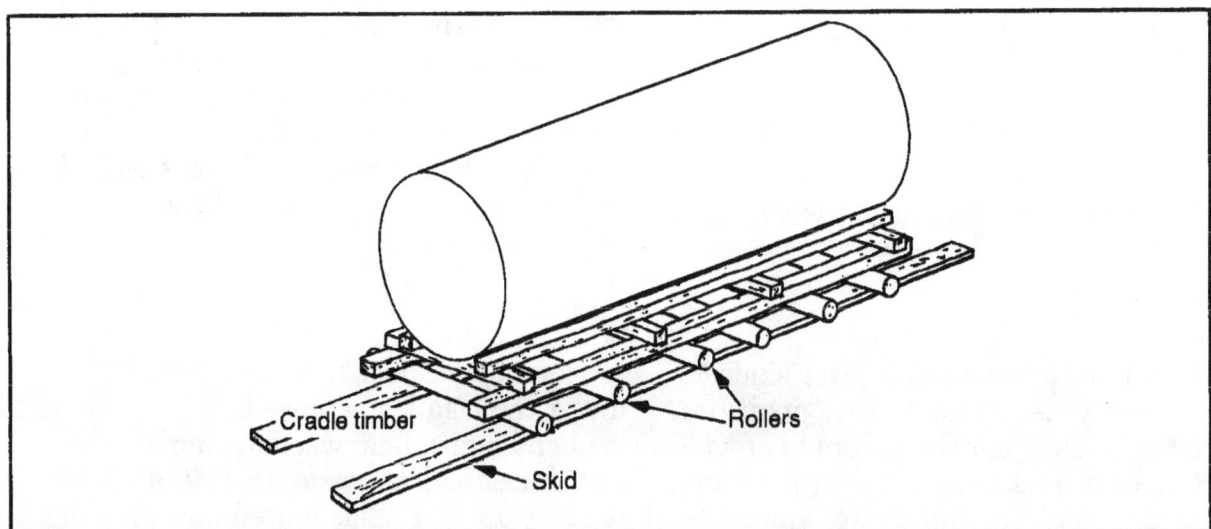

Figure 5-13. Using skids and rollers

the rollers must form continuous support. Ordinarily, place four to six rollers under the load to be moved (see *Figure 5-13, page 5-17*).

Place several rollers in front of the load and roll the load slowly forward onto the rollers. As the load passes, rollers are left clear behind the load and are picked up and placed in front of the load so that there is a continuous path of rollers. In making a turn with a load on rollers, incline the front rollers slightly in the direction of the turn and the rear rollers in the opposite direction. This inclination of the rollers may be made by striking them sharply with a sledge. For moving lighter loads, make up the rollers and set on axles in side beams as a semipermanent conveyor. Permanent metal roller conveyors are available (see*Figure 5-14*)They are usually made in sections.

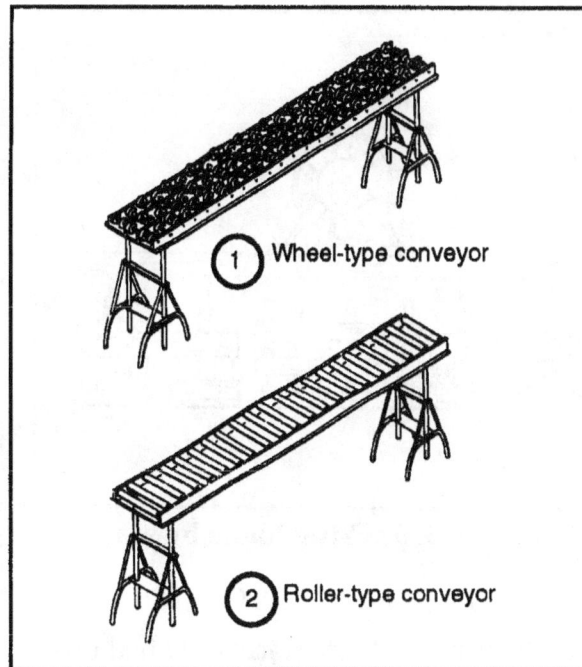

Figure 5-14. Metal conveyors

JACKS

To place cribbing, skids, or rollers, you may have to lift and lower the load for a short distance. Jacks are used for this purpose. Jacks are used also for precision placement of heavy loads, such as bridge spans. A number of different styles of jacks are available, but only use heavy duty hydraulic or screw-type jacks. The number of jacks used will depend on the weight of the load and the rated capacity of the jacks. Be certain that the jacks are provided with a solid footing, preferably wooden blocking. Cribbing is frequently used in lifting loads by jacking stages (see*Figure 5-15*)The procedure requires—

- Blocking the jacks.
- Raising the object to the maximum height of the jacks to permit cribbing to be put directly under the load.

- Lowering the load onto the cribbing.

Repeat this process as many times as necessary to lift the load to the desired height.

Jacks are available in capacities from 5 to 100 tons (see*Figure 5-16*)Small capacity jacks are operated through a rack bar or screw, while those of large capacity are usually operated hydraulically.

RATCHET-LEVER JACKS

The ratchet lever jack, available to engineer troops as part of panel bridge equipment, is a rack-bar jack that has a rated capacity of 15 tons (see*Figure 5-16, A*)It has a foot lift by which loads close to its base can be engaged. The foot capacity is 7 1/2 tons.

Figure 5-15. Jacking loads by stages

STEAMBOAT RATCHETS

Steamboat ratchets (sometimes called pushing-and-pulling jacks) are ratchet screw jacks of 10-ton rated capacity with end fittings that permit pulling parts together or pushing them apart (see *Figure 5-16, B).*

Their principal uses are for tightening lines or lashings and for spreading or bracing parts in bridge construction.

SCREW JACKS

Screw jacks have a rated capacity of 12 tons (see *Figure 5-16, C)* They are about 13 inches high when closed and have a safe rise of at least 7 inches. These jacks are issued with the pioneer set and can be used for general purposes, including steel erection.

HYDRAULIC JACKS

Hydraulic jacks are available in Class IV supplies in capacities up to 100 tons (see *Figure 5-16, D)* Loads normally encountered by engineer troops do not require large capacity hydralic jacks. Those supplied with the squad pioneer set are 11 inches high and have a rated capacity of 12 tons and a rise of at least 5 1/4 inches. They are large enough for usual construction needs.

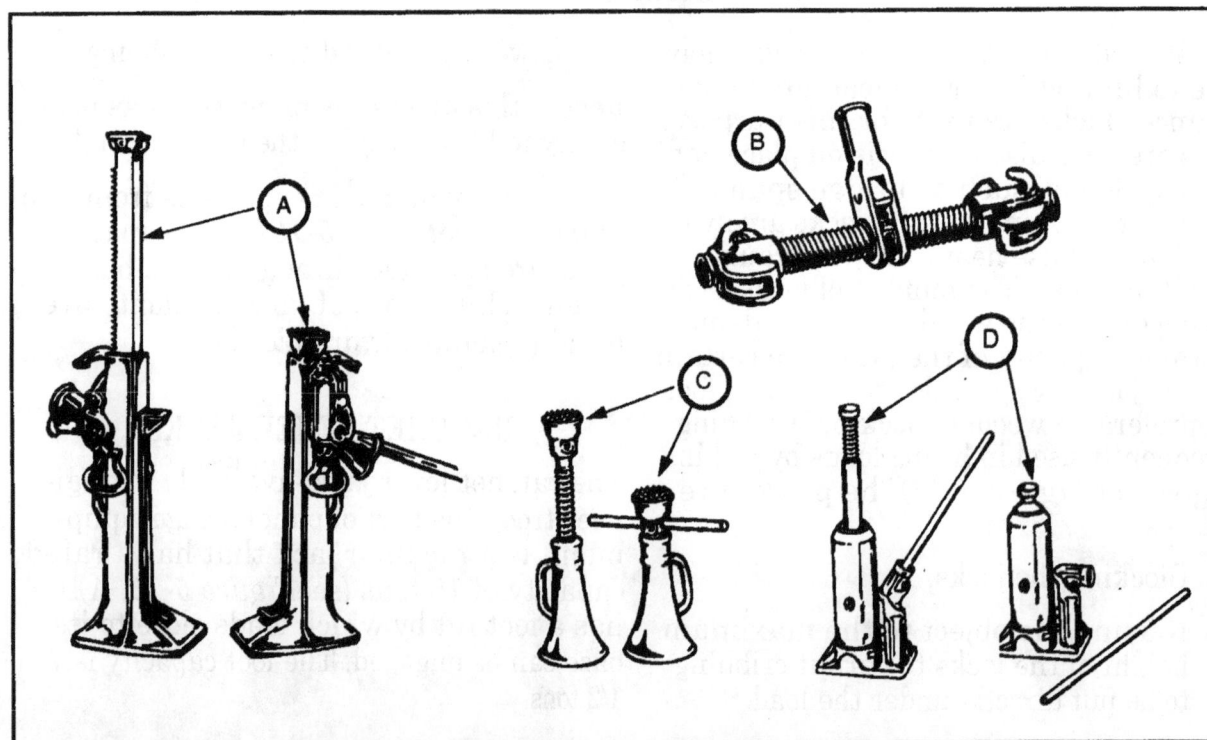

Figure 5-16. Mechanical and hydraulic jacks

CHAPTER 6

Scaffolds

Construction jobs may require several kinds of scaffolds to permit easy working procedures. Scaffolds may range from individual planks placed on structural members of the building to involved patent scaffolding. Scaffold planks are placed as a decking over—

- Swinging scaffolds.

- Suspended scaffolds.

- Needle-beam scaffolds.

- Double-pole, built-up, independent scaffolds.

Scaffold planks are of various sizes, including 2 inches by 9 inches by 13 feet, 2 inches by 10 inches by 16 feet, and 2 inches by 12 inches by 16 feet. You may need 3-inch-thick scaffold planks for platforms that must hold heavy loads or withstand movements. Planks with holes or splits are not suitable for scaffolding if the diameter of

the hole is more than 1 inch or if the split extends more than 3 inches in from the end. Use 3-inch planks to build the temporary floor used for constructing steel buildings because of the possibility that a heavy steel member might be rested temporarily on the planks. Lay single scaffold planks across beams of upper floors or roofs to form working areas or runways (see Figure 6-1, page 6-2). Run each plank from beam to beam, with not more than a few inches of any plank projecting beyond the end of the supporting beam. Overhangs are dangerous because people may step on them and overbalance the scaffold plank. When laying planking continuously, as in a runway, lay the planks so that their ends overlap. You can stagger single plank runs so that each plank is offset with reference to the next plank in the run. It is advisable to use two layers of planking on large working areas to increase the freedom of movement.

SWINGING SCAFFOLDS

The swinging, single plank, or platform type of scaffold must always be secured to the building or structure to prevent it from moving away and causing someone to fall. When swinging scaffolds are suspended adjacent to each other, planks should never be placed so as to form a bridge between them.

SINGLE-PLANK SWINGING SCAFFOLDS

A single scaffold plank maybe swung over the edge of a building with two ropes by using a scaffold hitch at each end (see Figures 6-2, page 6-2, and 2-28, page 2-20). tackle may be inserted in place of ropes for lowering and hoisting. This type of swinging scaffold is suitable for one person.

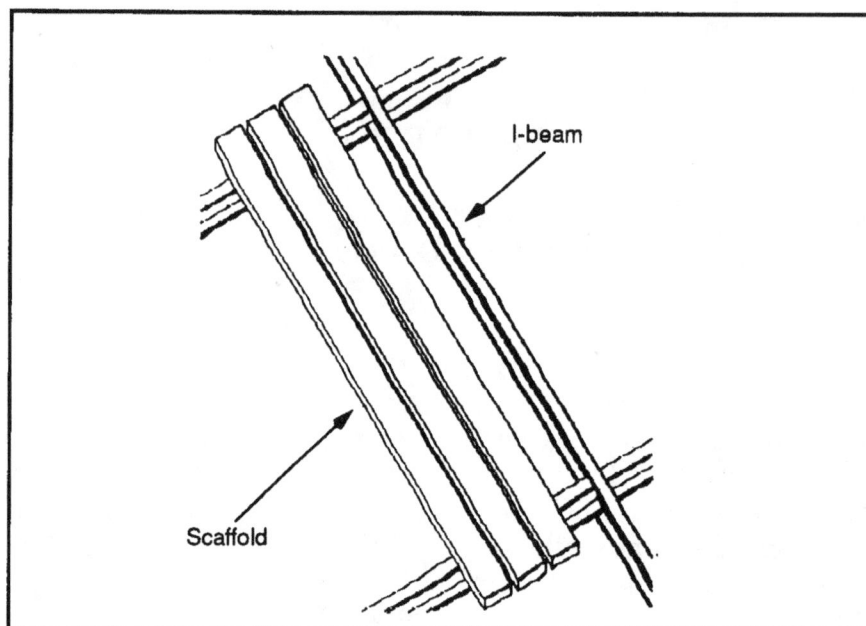

Figure 6-1. Scaffold planks in place

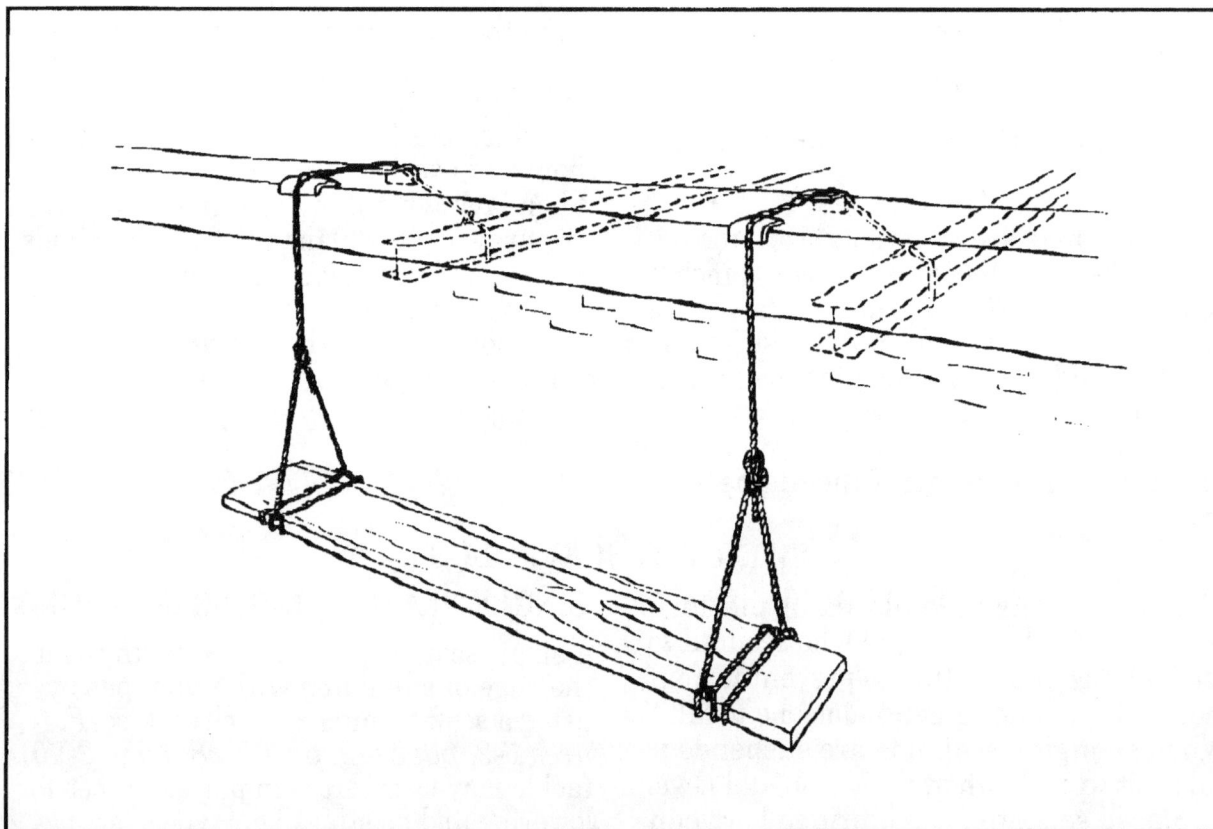

Figure 6-2. Single-plank swinging scaffold

SWINGING PLATFORM SCAFFOLDS

The swinging platform scaffold consists of a frame similar in appearance to a ladder with a decking of wood slats *(see Figure 6-3)*. It is supported near each end by a steel stirrup to which the lower block of a set of manila rope falls is attached. The scaffold is supported by hooks or anchors on the roof of a structure. The fall line of the tackle must be secured to a member of the scaffold when in final position to prevent it from falling.

SUSPENDED SCAFFOLDS

Suspended scaffolds are heavier than swinging scaffolds and are usually supported on outriggers at the roof. From each outrigger, cables lead to hand winches on the scaffold. This type of scaffold is raised or lowered by operating the hand winches, which must contain a locking device. The scaffold may be made up in almost any width up to about 6 feet and may be 12 feet long, depending on the size of the putlogs, or longitudinal supports, under the scaffold. A light roof may be included on this type of scaffold to protect people from falling debris.

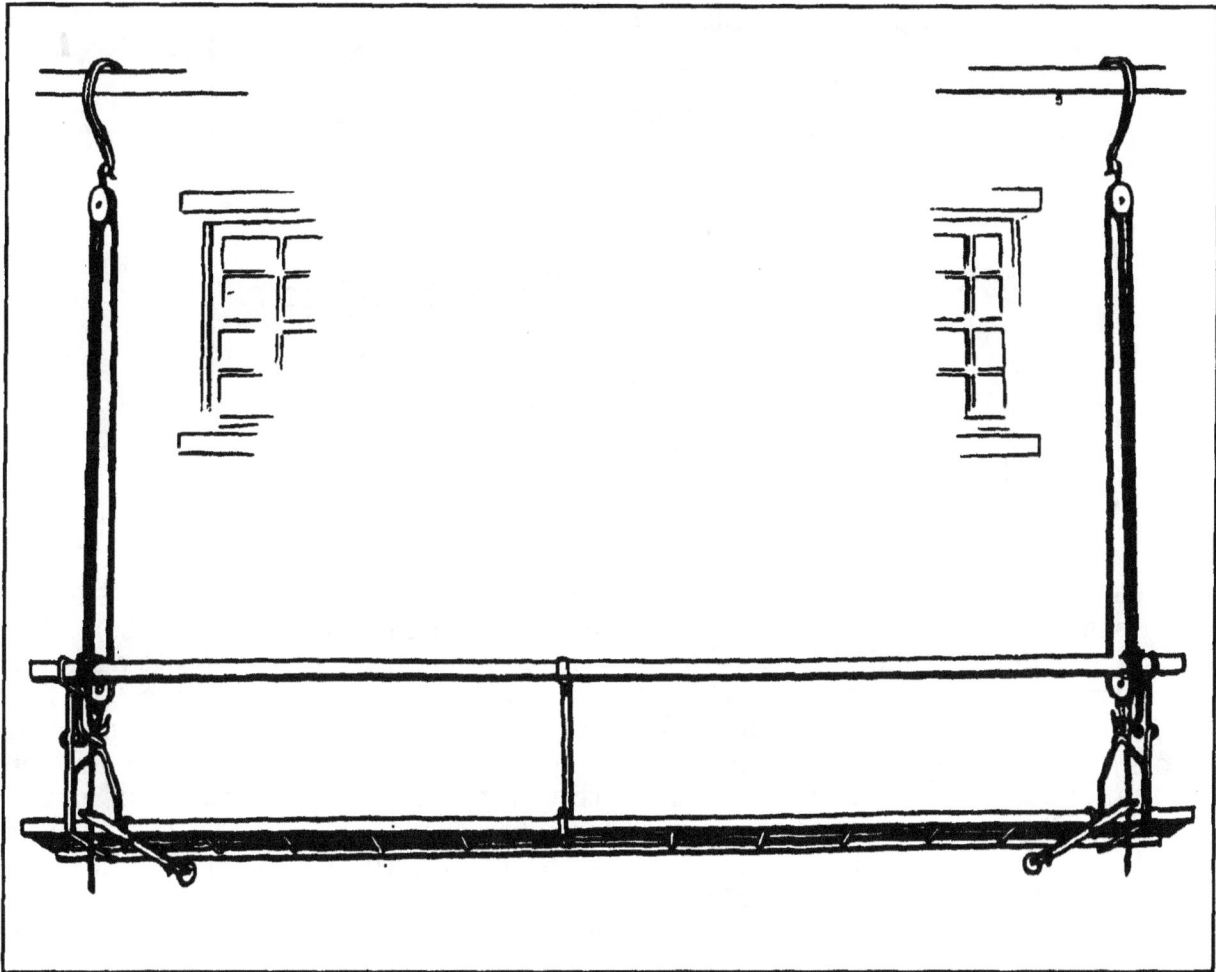

Figure 6-3. Swinging platform scaffold

NEEDLE-BEAM SCAFFOLDS

This type of scaffold is used only for temporary jobs. No material should be stored on this scaffold. In needle-beam scaffolding, two 4- by 6-inch, or similar size, timbers are suspended by ropes. A decking of 2-inch scaffold plank is placed across the needle beams, which should be placed about 10 feet apart. Needle-beam scaffolding is often used by riveting gangs working on steel structures because of the necessity for frequent changes of location and because of its adaptability to different situations(see

Figure 6-4)A scaffold hitch is used in the rope supporting the needle beams to prevent them from rolling or turning over(see Figure 2-28, page 2-20)The hanging lines are usually of 1 1/4-inch manila rope. The rope is hitched to the needle beam, carried up over a structural beam or other support, and then down again under the needle beam so the latter has a complete loop of rope under it. The rope is then passed over the support again and fastened around itself by two half hitches.

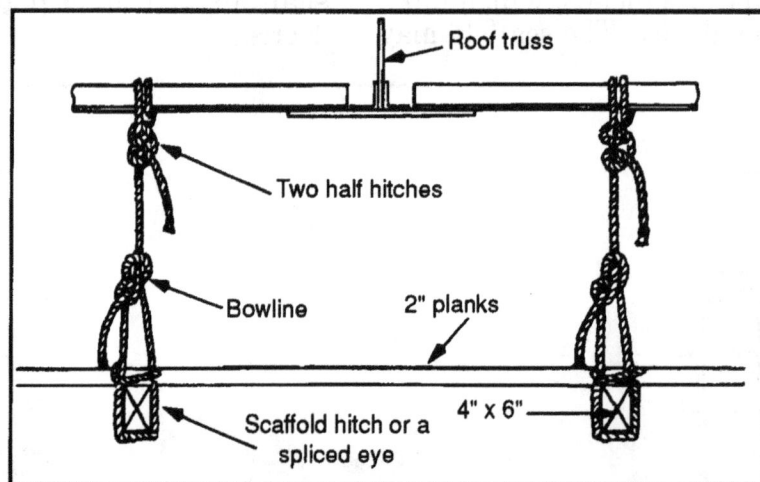

Figure 6-4. Needle-beam scaffold

DOUBLE-POLE BUILT-UP SCAFFOLDS

The double-pole built-up scaffold (steel or wood), sometimes called the independent scaffold, is completely independent of the main structure. Several types of patent independent scaffolding are available for simple and rapid erection(see Figure 6-5). The scaffolding can be built from wood, if necessary. The scaffold uprights are braced with diagonal members, and the working level is covered with a platform of planks. All bracing must form triangles. The base of each column requires adequate footing plates for the bearing area on the ground. Patented steel scaffolding is usually erected

by placing the two uprights on the ground and inserting the diagonal members. The diagonal members have end fittings that permit rapid locking-in position. The first tier is set on steel bases on the ground. A second tier is placed in the same manner on the first tier, with the bottom of each upright locked to the top of the lower tier. A third and fourth upright can be placed on the ground level and locked to the first set with diagonal bracing. The scaffolding can be built as high as desired, but high scaffolding should be tied in to the main structure.

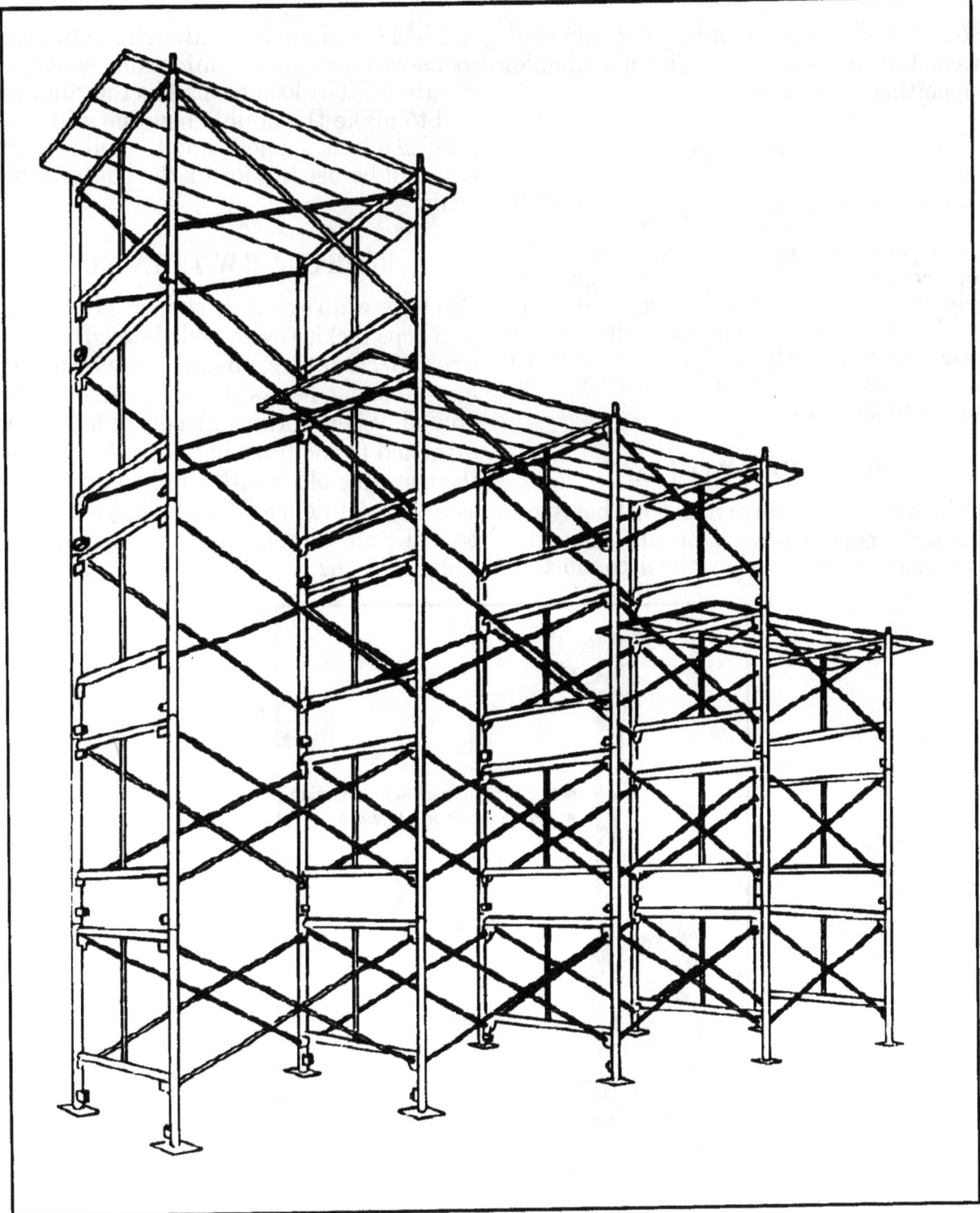

Figure 6-5. Independent scaffolding

BOATSWAIN'S CHAIRS

Boatswain's chairs can be made several ways, but they usually consist of a sling for supporting one person.

ROPE CHAIR

You can make a rope boatswain's chair by using a double bowline and a rolling hitch (see Figure 6-6).One person can operate the rope seat to lower himself by releasing the grip of the rolling hitch. A slight twist with the hand on the hitch permits the suspension line to slip through it, but when the hand pressure on the hitch is released, the hitch will hold firmly.

ROPE CHAIR WITH SEAT

If the rope boatswain's chair must be used to support a person at work for some time, the rope may cause considerable discomfort. A notched board inserted through the two leg loops will provide a comfortable seat (see Figure 6-7).The loop formed as the running end to make the double bowline will still provide a back support, and the rolling hitch can still be used to lower the boatswain's chair.

ROPE CHAIR WITH TACKLE

The boatswain's chair is supported by a four part rope tackle (two double blocks [see Figure 6-8]).One person can raise or lower himself or can be assisted by a person on the ground. When working alone, the fall line is attached to the lines between the seat and the traveling block with a rolling hitch. As a safety precaution, a figure-eight knot should be tied after the rolling hitch to prevent accidental untying.

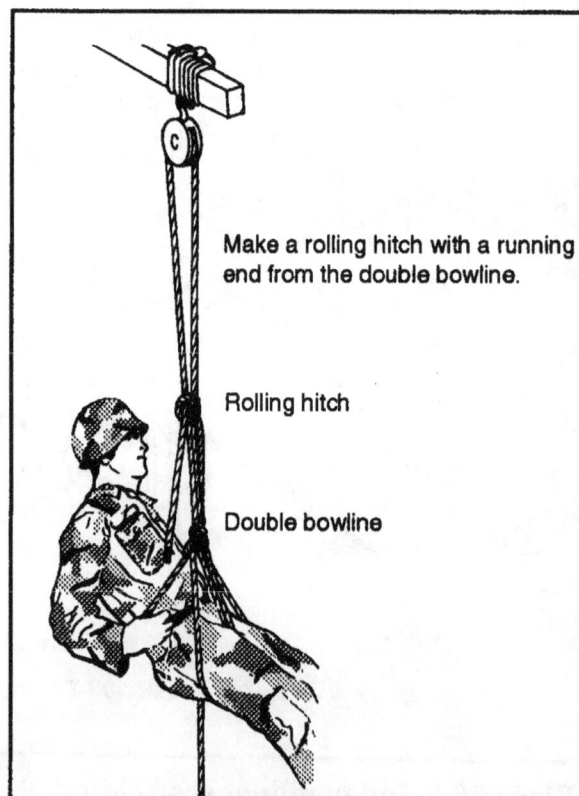

Make a rolling hitch with a running end from the double bowline.

Rolling hitch

Double bowline

Figure 6-6. Boatswain's chair

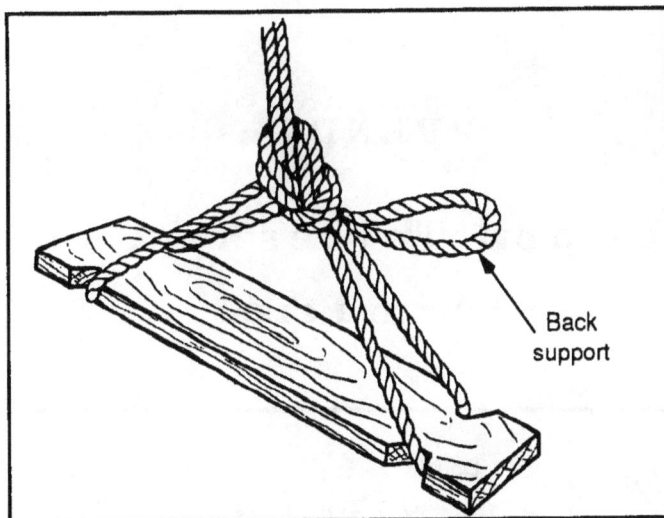

Figure 6-7. Boatswain's chair with seat

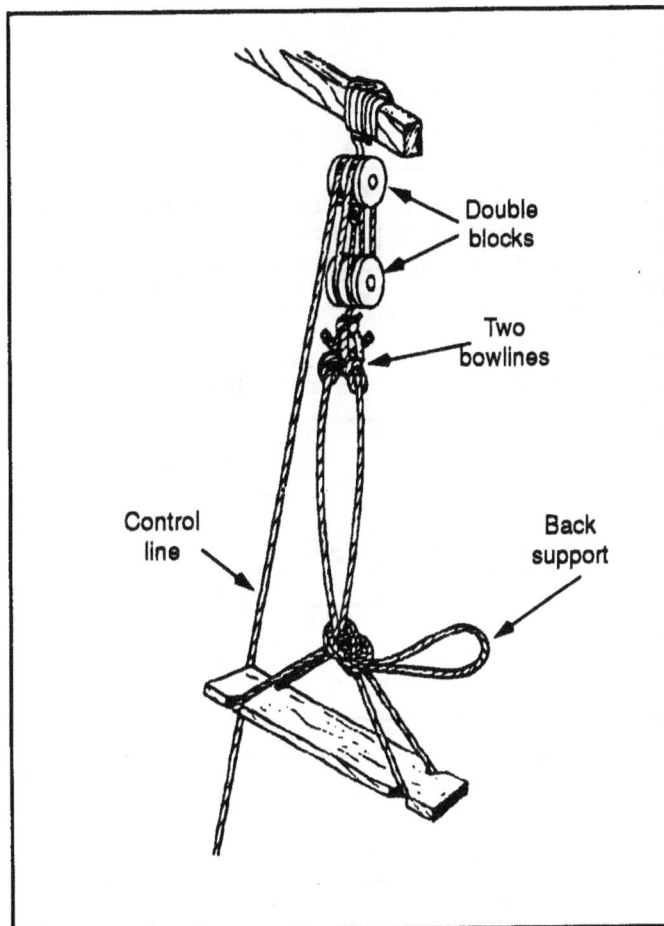

Figure 6-8. Boatswain's chair with tackle

APPENDIX

Figures and Tables of Useful Information

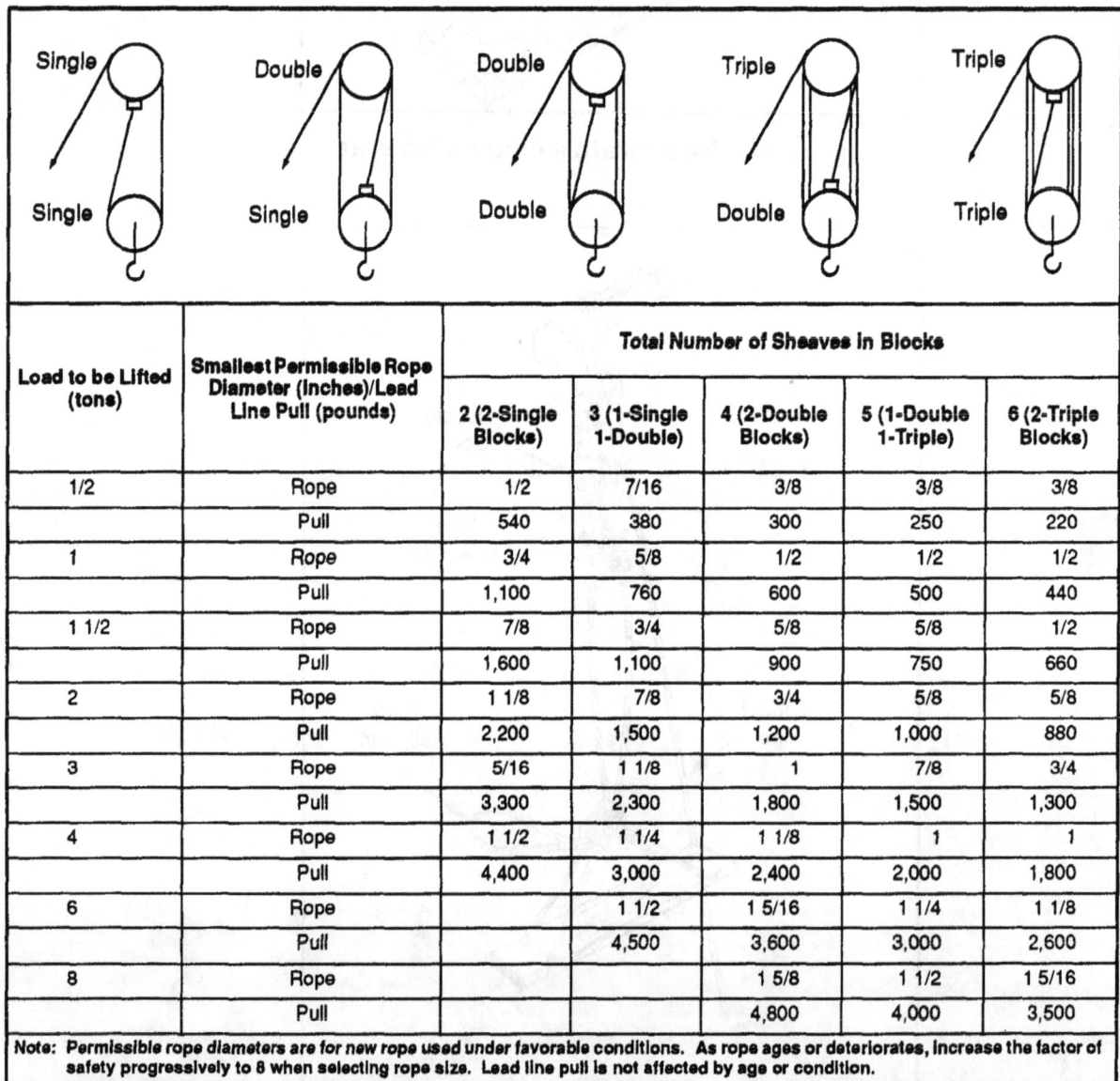

Load to be Lifted (tons)	Smallest Permissible Rope Diameter (Inches)/Lead Line Pull (pounds)	Total Number of Sheaves in Blocks				
		2 (2-Single Blocks)	3 (1-Single 1-Double)	4 (2-Double Blocks)	5 (1-Double 1-Triple)	6 (2-Triple Blocks)
1/2	Rope	1/2	7/16	3/8	3/8	3/8
	Pull	540	380	300	250	220
1	Rope	3/4	5/8	1/2	1/2	1/2
	Pull	1,100	760	600	500	440
1 1/2	Rope	7/8	3/4	5/8	5/8	1/2
	Pull	1,600	1,100	900	750	660
2	Rope	1 1/8	7/8	3/4	5/8	5/8
	Pull	2,200	1,500	1,200	1,000	880
3	Rope	5/16	1 1/8	1	7/8	3/4
	Pull	3,300	2,300	1,800	1,500	1,300
4	Rope	1 1/2	1 1/4	1 1/8	1	1
	Pull	4,400	3,000	2,400	2,000	1,800
6	Rope		1 1/2	1 5/16	1 1/4	1 1/8
	Pull		4,500	3,600	3,000	2,600
8	Rope			1 5/8	1 1/2	1 5/16
	Pull			4,800	4,000	3,500

Note: Permissible rope diameters are for new rope used under favorable conditions. As rope ages or deteriorates, increase the factor of safety progressively to 8 when selecting rope size. Lead line pull is not affected by age or condition.

Figure A-1. Simple block-and-tackle rigging for manila rope (FS 3)

Table A-1. Simple block and tackle rigging for plow steel wire rope (FS 6)

Load to be Lifted (tons)	Smallest Permissible Rope Diameter (Inches)/ Lead Line Pull (pounds)	Total Number of Sheaves in Blocks				
		2 (2 Single Blocks)	3 (1-Single 1-Double)	4 (2-Double Blocks)	5 (1-Double 1-Triple)	6 (2-Triple Blocks)
1	Rope	3/8	3/8	3/8	3/8	3/8
	Pull	1,000	720	560	460	400
2	Rope	1/2	3/8	3/8	3/8	3/8
	Pull	2,100	1,400	1,100	920	800
4	Rope	5/8	1/2	1/2	3/8	3/8
	Pull	4,200	2,900	2,200	1,800	1,600
6	Rope	3/4	5/8	5/8	1/2	1/2
	Pull	6,200	4,300	3,400	2,800	2,400
8	Rope	7/8	3/4	5/8	5/8	5/8
	Pull	8,300	5,800	4,500	3,700	3,200
10	Rope	1	7/8	3/4	5/8	5/8
	Pull	10,400	7,200	5,600	4,600	4,000
15	Rope	1 1/8	1	7/8	3/4	3/4
	Pull	15,600	10,800	8,400	6,900	6,000
20	Rope	1 1/2	1 1/8	1	7/8	7/8
	Pull	20,800	14,400	11,200	9,200	8,000

Table A-2. Recommended sizes of tackle blocks

Wire Rope		Manila Rope	
Rope Diameter (Inches)	Outside Diameter of Sheave (Inches)	Rope Diameter (Inches)	Length of Shell (Inches)
3/8	6-8	1/2	4
1/2	8-10	5/8	6
5/8	10-12	3/4	6-7
3/4	12-16	7/8	7-8
7/8	14-18	1	8-10
1	14-20	1 1/8	8-10
		1 1/4	10-12
		1 1/2	12-14
		1 3/4	14-16

Note: Largest diameter of sheave for a given size of rope is preferred, when available, except that for 6 x 37 wire rope, the smaller diameter of sheave is suitable.

Table A-3. Bearing capacity of soils

General Description	Condition	Safe Allowable Pressure (PSI)
Fine-grained soils: clays, silts, very fine sands, or mixtures of these containing few coarse particles of sand or gravel. Classification: MH, CH, OH, ML, CL, and OL.	Soft, unconsolidated, having high moisture content (mud)	1,000
	Stiff, partly consolidated, medium moisture content	4,000
	Hard, well consolidated, low moisture content (slightly damp to dry)	8,000
Sands and well-graded sandy soils, containing some silt and clay. Classification: SW, SC, SP, and SF.	Loose, not confined	3,000
	Loose, confined	5,000
	Compact	10,000
Gravel and well-graded gravelly soils containing some sand, silt and clay. Classification: GW, GC, and GP.	Loose, not confined	4,000
	Loose, confined	6,000
	Compact	12,000
	Cemented sand and gravel	16,000
Rock	Poor quality rock, soft and fractured; also hardpan	10,000
	Good quality; hard and solid	20,000

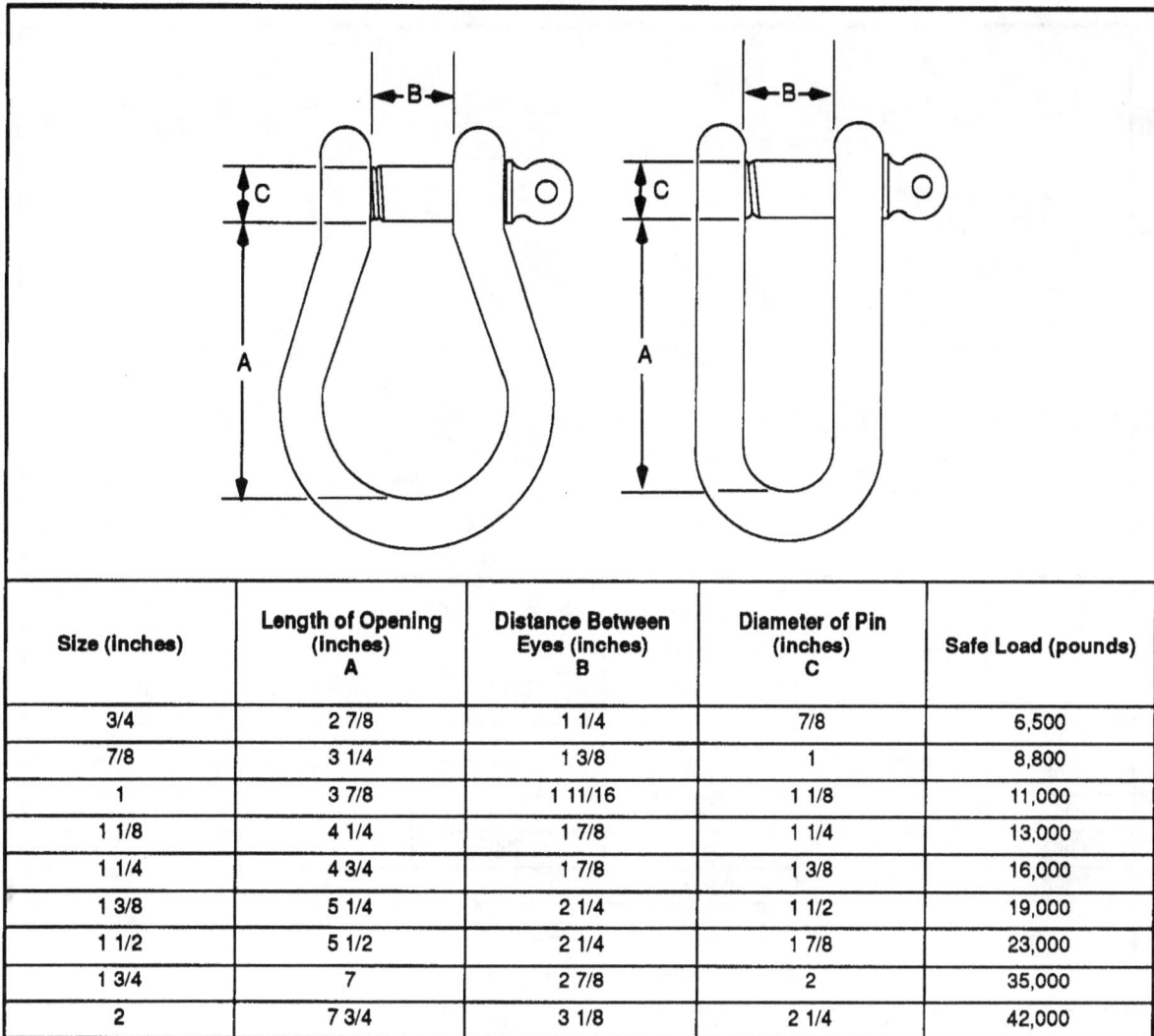

Size (inches)	Length of Opening (inches) A	Distance Between Eyes (inches) B	Diameter of Pin (inches) C	Safe Load (pounds)
3/4	2 7/8	1 1/4	7/8	6,500
7/8	3 1/4	1 3/8	1	8,800
1	3 7/8	1 11/16	1 1/8	11,000
1 1/8	4 1/4	1 7/8	1 1/4	13,000
1 1/4	4 3/4	1 7/8	1 3/8	16,000
1 3/8	5 1/4	2 1/4	1 1/2	19,000
1 1/2	5 1/2	2 1/4	1 7/8	23,000
1 3/4	7	2 7/8	2	35,000
2	7 3/4	3 1/8	2 1/4	42,000

Figure A-2. Safe loads on screw-pin shackles

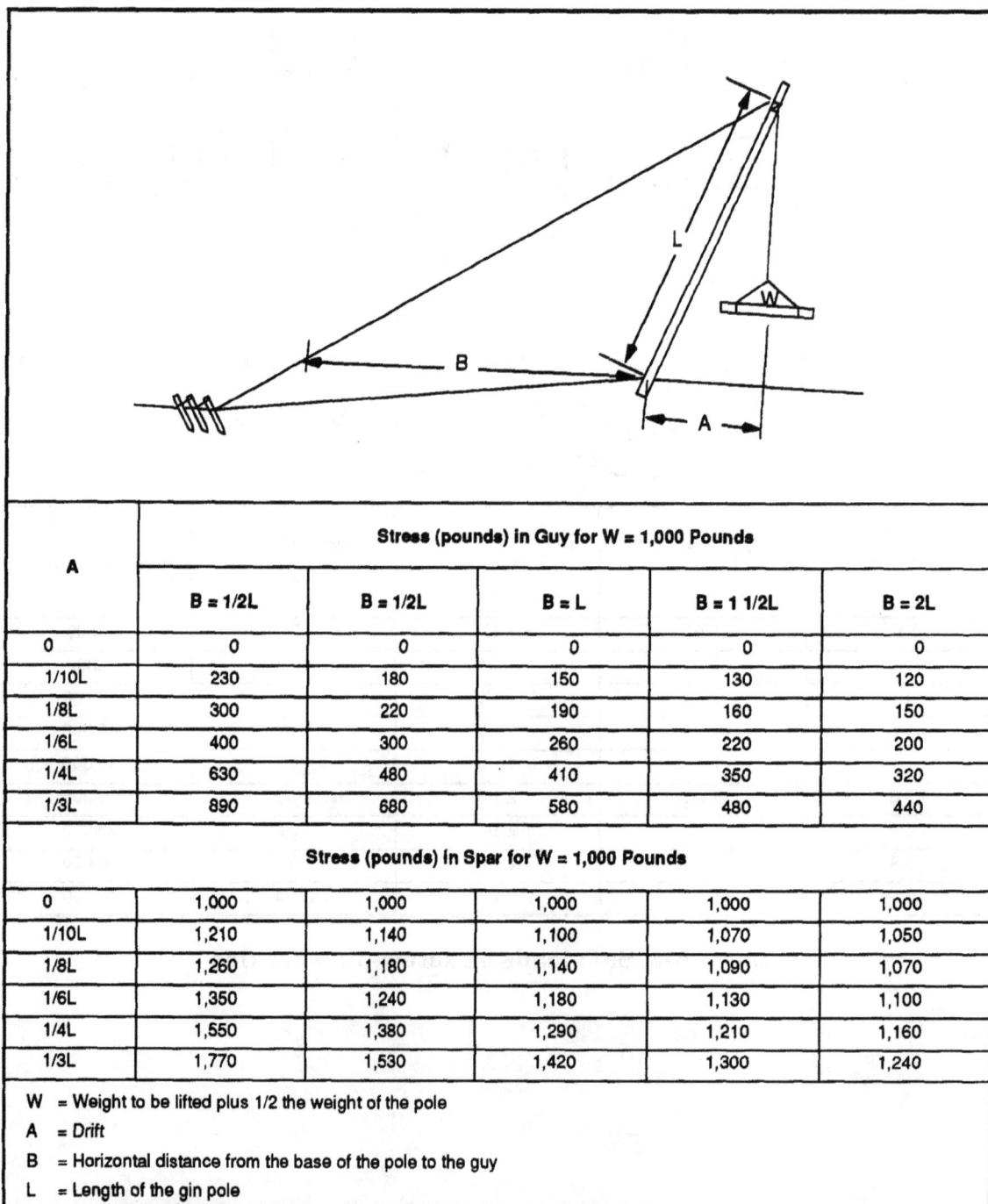

A	Stress (pounds) in Guy for W = 1,000 Pounds				
	B = 1/2L	B = 1/2L	B = L	B = 1 1/2L	B = 2L
0	0	0	0	0	0
1/10L	230	180	150	130	120
1/8L	300	220	190	160	150
1/6L	400	300	260	220	200
1/4L	630	480	410	350	320
1/3L	890	680	580	480	440
Stress (pounds) in Spar for W = 1,000 Pounds					
0	1,000	1,000	1,000	1,000	1,000
1/10L	1,210	1,140	1,100	1,070	1,050
1/8L	1,260	1,180	1,140	1,090	1,070
1/6L	1,350	1,240	1,180	1,130	1,100
1/4L	1,550	1,380	1,290	1,210	1,160
1/3L	1,770	1,530	1,420	1,300	1,240

W = Weight to be lifted plus 1/2 the weight of the pole
A = Drift
B = Horizontal distance from the base of the pole to the guy
L = Length of the gin pole

Figure A-3. Stresses in guys and spars of gin poles

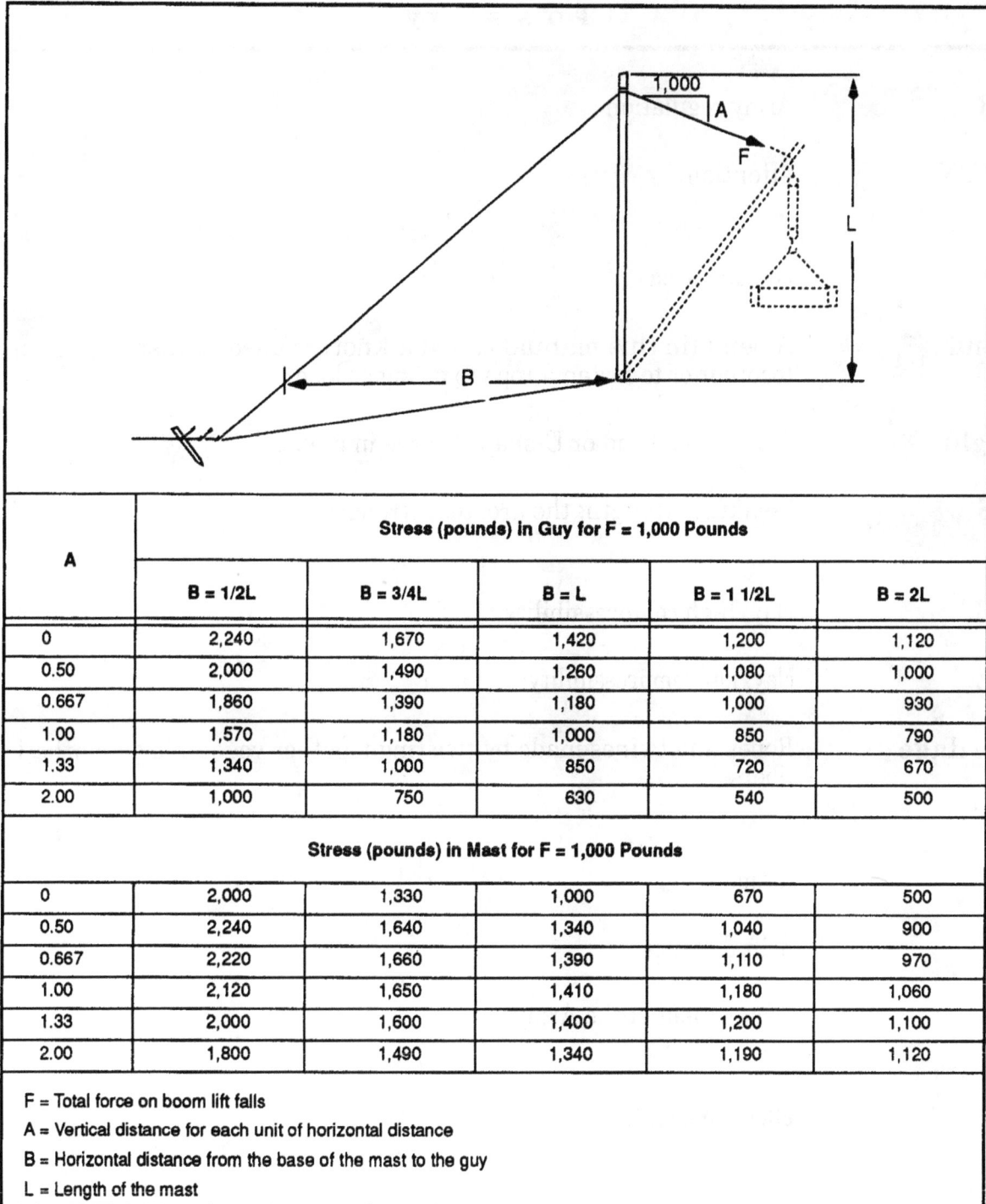

A	Stress (pounds) in Guy for F = 1,000 Pounds				
	B = 1/2L	B = 3/4L	B = L	B = 1 1/2L	B = 2L
0	2,240	1,670	1,420	1,200	1,120
0.50	2,000	1,490	1,260	1,080	1,000
0.667	1,860	1,390	1,180	1,000	930
1.00	1,570	1,180	1,000	850	790
1.33	1,340	1,000	850	720	670
2.00	1,000	750	630	540	500
	Stress (pounds) in Mast for F = 1,000 Pounds				
0	2,000	1,330	1,000	670	500
0.50	2,240	1,640	1,340	1,040	900
0.667	2,220	1,660	1,390	1,110	970
1.00	2,120	1,650	1,410	1,180	1,060
1.33	2,000	1,600	1,400	1,200	1,100
2.00	1,800	1,490	1,340	1,190	1,120

F = Total force on boom lift falls
A = Vertical distance for each unit of horizontal distance
B = Horizontal distance from the base of the mast to the guy
L = Length of the mast

Figure A-4. Stresses in guys and mast of guy derrick

Glossary

AR   Army regulation

ATTN   attention

BA   bearing area

bend   A bend (in this manual called a knot) is used to fasten two ropes together or to fasten a rope to a ring or loop.

bight   A bight is a bend or U-shaped curve in a rope.

BS   breaking strength; the greatest strength.

CH   clay, high compressibility

CL   clay, low compressibility

cordage   Ropes and twines made by twisting together vegetable or synthetic fibers.

D   diameter

D   drift distance

DA   Department of the Army

EL   effective length

ENG   engineer

FM   field manual

FS   factor of safety

GC	clayey gravel
GP	poorly graded gravel
GW	well-graded gravel
HD	horizontal distance
HP	holding power
HQ	headquarters
IPS	improved plow steel
L	length of the sling
line	A line (sometimes called a rope) is a thread, string, cord, or rope, especially a comparatively slender and strong cord. This manual will use the word rope rather than line in describing knots, hitches, rigging, and the like.
loop	A loop is formed by crossing the running end cover or under the standing part, forming a ring or circle in the rope.
L/d	length-to-diameter ratio
L/t	length-to-thickness ratio
MA	mechanical advantage
MD	mean depth
MH	silt, high compressibility
ML	silt, low compressibility
MPS	mild plow steel

N   number of slings

No   number

OH   organic soil, high compressibility

OL   organic soil, low compressibility

overhand turn or loop An overhand turn or loop is made when the running end passes
                        over the standing part.

PS   plow steel

psi   pound(s) per square inch

rope   A rope (often called a line) is a large, stout cord made of strands of
                        fiber or wire that are twisted or braided together.

round turn   A round turn is a modified turn, but with the running end leaving the
                        circle in the same general direction as the standing part.

running end   The running end is the free or working end of a rope.

SC   clayey sandy soil

SF   finely graded sand

SP   poorly graded sand

SR   slope ratio

standing part   The standing part is the rest of the rope, excluding the running end.

SW   well-graded sand

SWC   safe working capacity

T	tension
TB	technical bulletin
TC	training circular
TL	timber length
TM	training manual
TRADOC	United States Army Training and Doctrine Command
turn	A turn is the placing of a loop around a specific object (such as a post, rail, or ring) with the running end continuing in a direction opposite to the standing part.
underhand turn or loop	An underhand turn or loop is made when the running end passes under the standing part.
US	United States (of America)
V	vertical distance
VD	vertical distance
W	weight of the load to be lifted
W3	width of spar(s)
WL	width of the load
WST	width of the sloping trench
Y	Perpendicular distance from the rear guy line to the base of the gin pole or, for shears, to a point on the ground midway between the shears legs.

R e f e r e n c e s

SOURCES USED

These are the sources quoted or paraphrased in this publication.

Army Publications

Army Regulations (ARs)

AR 59-3. *Air Transportation Movement of Cargo by Scheduled Military and Commercial Air Transportation -- CONUS Outbound.* 1 February 1981.

Field Manuals (FMs)

FM 5-34. *Engineer Field Data.* 14 September 1987.

FM 5-434. *Earthmoving Operations.* 30 September 1992,

FM 10-500-7. *Airdrop Derigging and Recovery Procedures.* 20 September 1994.

FM 20-22. *Vehicle Recovery Operations (FMFRP 4-19).* 18 September 1990.

FM 55-9. *Unit Air Movement Planning.* 5 April 1993.

FM 55-12. *Movement of Units in Air Force Aircraft (AFM 76-7; FMFM 4-6; OPNAVINST 4630.27A).* 10 November 1989,

FM 55-15. *Transportation Reference Data.* 9 June 1986.

Supply Catalog (SC)

SC 5180-90-CL-N17. *Tool Kit Rigging, Wire Rope: Cutting, Clamping, and Splicing w/Chest.* 23 October 1981.

Technical Bulletins (TBs)

TB 43-0142. *Safety Inspection and Testing of Lifting Devices.* 30 August 1993.

TB ENG 317. *Air Movement Instructions: Grouping, Modification, Disassembly and Reassembly for Crane, Shovel, Truck Mounted, 20 Ton, 3/4 Cubic Yard, Gasoline Driven, Garwood Model M-20-B.* 28 June 1962.

TB ENG 324. *Air Movement Instructions (Grouping, Modification, Disassembly, and Reassembly) for Mixer, Concrete, GED, Trailer Mounted (Construction Machinery Model 16S).* 2 July 1962.

TB ENG 326. *Air Movement Instructions: (Grouping, Modification, Disassembly, and Reassembly) for Scraper, Earth Moving, Towed, 12 Cubic Yard, Cable Operated, Letorneau-Westinghouse Model LPO.* 9 July 1962.

TB ENG 330,*Air Movement Instructions (Grouping, Modification, Disassembly, and Reassembly) for Truck, Stake: 5-Ton, 6x6; Military Bridging on ORD M-139 Chassis.* 3 July 1962,

Training Circular (TC)

TC 90-6-1,*Military Mountaineering.* 26 April 1989.

Technical Manuals (TMs)

TM 5-270,*Cableways, Tramways, and Suspension Bridges.* 21 May 1964.

TM 10-500-70,*Airdrop of Supplies and Equipment: Rigging Dry Bulk Materials and Potable Water for Free Drop.* 2 November 1967.

DOCUMENTS NEEDED

These documents must be available to the intended users of this publication.

Department of the Army (DA) Forms

DA Form 2028.*Recommended Changes to Publications and Blank Forms.* February 1974.

Index

PIN: 074097-000

www.ingramcontent.com/pod-product-compliance
Lightning Source LLC
Chambersburg PA
CBHW080419030426
42335CB00020B/2510